Occupational Accident Research

HD
7262.
5.
S8

Occupational Accident Research

Proceedings of the International Seminar on Occupational Accident Research,
Saltsjöbaden, Sweden, 5—9 September 1983

Edited by
Urban Kjellén

Occupational Accident Research Unit, Royal Institute of Technology, S 100 44 Stockholm, Sweden

This set of papers has been published as a special issue of Journal of Occupational Accidents, Vol. 6, Nos. 1—3

ELSEVIER, Amsterdam—Oxford—New York—Tokyo   1984

ELSEVIER SCIENCE PUBLISHERS B.V.
Molenwerf 1,
P.O. Box 211, 1000 AE Amsterdam, The Netherlands

Distributors for the United States and Canada:

ELSEVIER SCIENCE PUBLISHING COMPANY INC.
52, Vanderbilt Avenue
New York, NY 10017, U.S.A.

ISBN 0-444-42415-6

© Elsevier Science Publishers B.V., 1984

All rights reserved. No part of this publication may be reproduced, stored in a retrieval system or transmitted in any form or by any means, electronic, mechanical, photocopying, recording or otherwise, without the prior written permission of the publisher, Elsevier Science Publishers B.V./Science & Technology Division, P.O. Box 330, 1000 AH Amsterdam, The Netherlands.

Special regulations for readers in the USA — This publication has been registered with the Copyright Clearance Center Inc. (CCC), Salem Massachusetts. Information can be obtained from the CCC about conditions under which photocopies of parts of this publication may be made in the USA. All other copyright questions, including photocopying outside of the USA, should be referred to the copyright owner, Elsevier Science Publishers B.V., unless otherwise specified.

Printed in The Netherlands

PREFACE

This book is a collection of papers and abstracts of reports to the International Seminar on Occupational Accident Research held at Saltsjöbaden, Sweden, on September 5—9, 1983.

The seminar was organized by the Swedish Work Environment Fund, together with the Occupational Accident Research Unit of the Royal Institute of Technology in Stockholm. The terms of reference of the Swedish Work Environment Fund are to support research and development work which could contribute to creating more humane working conditions in Sweden. The Occupational Accident Research Unit was established in 1978 to carry out research, development and information activities within the area of occupational accidents.

The purpose of the seminar was to provide an opportunity to exchange experience and knowledge and to stimulate communication between scientists in the field of occupational accidents. A more specific purpose was to evaluate different theoretical and methodological approaches in accident research.

The seminar consisted of an introductory session, five special sessions and a closing session. The themes of the special sessions were selected from the results of a survey directed at Swedish accident researchers. They cover a variety of basic theoretical and methodological questions in accident research; i.e. evaluation of accident prevention strategies, validity and utility of theories and models of accidents, research application of occupational injury data systems, research strategies and methods applied to fall accidents, and safety effects of new production technology. These themes were addressed in approximately 50 presentations by invited lecturers and other participants. They represent various basic disciplines and are active in the study of occupational accidents as well as traffic, home and leisure-time accidents.

Accidents at work are still a serious problem. Severe accidents have become less common in Sweden, but the general accident rate has remained stable during the last decade. Substantial efforts to reduce and eliminate the effects of accidents have been made and must continue. Research is a valuable means of finding out directions for future efforts in accident prevention.

Many studies carried out within the field of occupational accidents aim at solving specific and urgent accident problems. Within the scientific community there is an increasing awareness of the need for a theoretical and methodological development in order to make the generalization of results from single studies possible. An international exchange of information and experience is important in this work, especially for a small country like Sweden.

The reports to the seminar are published with the aim of widening and intensifying discussions of the development and future directions of occupational accident research.

We are indebted to many individuals for any success achieved by the seminar and this publication. First of all, we gratefully acknowledge the contributions of the lecturers and other participants. They were most cooperative in following our guidelines regarding preprints before the presentations and the revision of the reports for publication. We are very grateful to the members of the organizing committee and the organizers of the special sessions for their important roles in holding the seminar and producing this publication. The efforts of the session chairmen are also appreciated. Members of the staff of the Swedish Work Environment Fund were of invaluable assistance.

Urban Kjellén
Occupational Accident Research Unit

Bo Oscarsson
Swedish Work Environment Fund

CONTENTS

Preface
   U. Kjellén and B. Oscarsson..................................... v

INTRODUCTION

Future trends in accident research in European countries
   W.T. Singleton.. 3
Abstract
Setting priorities for occupational health and safety research
   Lynn Stewart Hewitt and Judith E. Maki Evans 13

EVALUATION OF ACCIDENT PREVENTION STRATEGIES
— METHODOLOGY AND PRACTICAL RESULTS

Is safety training worthwhile?
   A.R. Hale 17
A review of the traffic safety situation in Sweden with regard to
   different strategies and methods of evaluating traffic safety
   measures
   Göran Nilsson 35
Behavioural control through piece-rate wages
   Carin Sundström-Frisk 49
Psychological safety diagnosis
   U. Bernhardt, C. Graf Hoyos and G. Hauke...................... 61
Abstracts
Experience of implementing safety information and management
   systems in industrial companies
   M.T. Ho... 71
A behavioral approach to work motivation
   Judith L. Komaki 71
Alcohol and fatal work accidents
   Elisabeth Lagerlöf, Milan Valverius and Peter Westerholm 72
A problem-oriented, interdisciplinary approach to safety problems
   John Stoop 72
Interviewing employees about near accidents as a means of initiating
   accident prevention activities: A review of Swedish research
   Ned Carter... 73
Developing routines in efforts to prevent occupational accidents:
   An accident investigation group
   Ewa Menckel... 73

VALIDITY AND UTILITY OF THEORIES AND MODELS OF ACCIDENTS

Occupational accident research and systems approach
   Jacques Leplat. 77
Accidents, and disturbances in the flow of information
   Jorma Saari. 91
Application of human error analysis to occupational accident
   research
   W.T. Singleton. 107
The role of deviations in accident causation and control
   Urban Kjellén 117

Abstracts

Accident models: How underlying differences affect workplace
   safety
   Ludwig Benner, Jr. 127
Serious occupational injuries with special regard to the lack
   of risk control
   Jan Thorson 128
Use of case reports from work accidents
   Jens Rasmussen. 129
Epidemiology of occupational accidents
   Börje Bengtsson. 129
On hand injuries
   Tore J. Larsson 129
A tentative conceptual analysis of safety activity
   Ned Carter. 130
Validity and utility of experimental research on severe work
   vehicle accidents
   Lennart Strandberg. 130
Worker safety among wage-earners — the connection between
   individual factors, working conditions and industrial accidents
   Eila Riikonen 131
Hazards in stationary grinding machines
   Risto Tuominen and Markku Mattila. 131

RESEARCH APPLICATION OF OCCUPATIONAL INJURY DATA SYSTEMS AT NATIONAL OR BRANCH LEVEL

Descriptive epidemiology in job injury surveillance
   Patrick J. Coleman 135
Use of census data combined with occupational accident data
   Elisabet Broberg 147
Hand injuries in Sweden in 1980
   Annika Carlsson 155

Estimation of potential seriousness of accidents and near accidents
Heikki Laitinen 167
Abstracts
10 years experience of accident registration
Arne Rasmussen 175
How to use the information system on occupational injuries (ISA) in research
Jan Carlsson 175
Occupational accident data and safety research in Finland
Pekka Maijala 176
The analysis of injuries amongst workers in a hospital for chronic patients
Monique Lortie 176
Injury information systems for management
Derek Viner 177

RESEARCH STRATEGIES AND METHODS APPLIED TO FALL ACCIDENTS

Fall accidents at work, in the home and during leisure activities
Johan Lund 181
Abstracts
Advantages and limitations of various methods used to study occupational fall accident patterns
H. Harvey Cohen 195
The accident model applied to back injuries
J.D.G. Troup 195
Accident analysis, biomechanics, and tribology for slipping and falling injury prevention
Lennart Strandberg 196
Some aspects on mechanical testing of fall safety devices
H. Andersson 196
Research in fall protection at Ontario Hydro
Andrew C. Sulowski 196
Accidents involving falls from roofs — survey and technical preventive measures
Per-Olof Axelsson 197

SAFETY EFFECTS OF NEW PRODUCTION TECHNOLOGY

A statistical study of control systems and accidents at work
T. Backström and L. Harms-Ringdahl 201
Abstracts
Human aspects of safety in offshore maintenance
Reidar Østvik 211

Potentials and limitations of risk and safety analysis — experiences
from the SCRATCH program
  Jann H. Langseth...211
Identification of accident risks in maintenance
  Juoko Suokas..212
Software safety in microprocessor-based machinery
  Søren Lindskov Hansen...213
Practical utilization of safety analysis results
  J.R. Taylor...213
The worker as a safety resource in modern production systems
  J. Hovden and T. Sten...214
Accidents and variance control
  Gordon Robinson...214
Safety considerations in the design of factories — a study of three
cases
  G. Fång and L. Harms-Ringdahl...................................214

CONCLUSIONS

Occupational accident research: where have we been and where
are we going?
  Jerry L. Purswell and Kåre Rumar................................219

INTRODUCTION

FUTURE TRENDS IN ACCIDENT RESEARCH IN EUROPEAN COUNTRIES

W.T. SINGLETON

Applied Psychology Department, The University of Aston in Birmingham, College House, Gosta Green, Birmingham B4 7ET (United Kingdom)

ABSTRACT

Singleton, W.T., 1984. Future trends in accident research in European countries. *Journal of Occupational Accidents*, 6: 3—12.

In all European countries there has been extensive growth in accident studies since about 1970. This was stimulated and was also made possible by the increasing affluence of the period. Most of this work was practical and pragmatic with an emphasis on the collection and collation of data from real situations. The inevitable result has been increased awareness of the industrial situation but a dearth of new theoretical concepts. Most current theories are of the "systems" type which can explain anything but predict nothing.

For the remainder of this century studies are likely to be much more tightly controlled and assessed in cost/benefit terms with an emphasis on the epidemiological approach of monitoring the statistical situation to detect new hazards. The research emphasis will probably switch from accidents to the occupational diseases consequent upon the vast increase in the use of new chemicals in industry.

INTRODUCTION

Although the focus of interest is occupational accidents, the history and progress of accident research need not be studied too restrictively. A theory or a method which is developed for other issues such as transport or home accidents should be applicable also in the occupational field. For example, the studies of accident proneness have progressed in relation to industrial accidents (Greenwood and Woods, 1919) and transport accidents (Cresswell and Froggatt, 1963). The fact that these two studies using similar methodology are separated by more than forty years illustrates the slow rate of progress up to about 1970. The latter study was on bus drivers who exemplify the fundamental problem of accident statistics — categorisation. Where should accidents of bus drivers and other working drivers be placed if one wishes to separate occupational accidents from transport accidents? In the U.K. for example, road traffic accidents involving lorry drivers are not included in "accidents to employees at work". Since lorry driver is the most common male occupation and the job is dangerous they obviously make a considerable difference to any category where they are included or excluded. Changes in policy decisions or regulations can also distort trends in data. Again taking the U.K. as an example, self-employed persons and other non-employees were incorporated in occupational accidents in January 1981, thus inflating the data but, following social security scheme changes in April 1983, there will be a considerable fall in reported injuries at work. Nevertheless the examination of statistical data is a necessary starting point for the study of trends. In most advanced countries there were considerable changes in accident research activities around 1970. In Sweden in 1970 the Work Environment Commission was appointed, in the U.K. in 1970 the Robens Committee on Safety and Health at Work was appointed. In 1972, the Swedish Work Environment Fund was set up. In 1974 in Sweden, the National Board of Occupational Safety and Health and the Labour Inspectorate were reorganised. In 1974 in the U.K. the new Health and Safety at Work Act became law. This set up the Health and Safety Executive and included a complete reorganisation and integration of previously separate inspectorates. Similar events were occurring in other countries at the time. In Germany, in 1971, approval was given for the transfer of the Federal Institute for Safety at Work from Koblenz to Dortmund with greatly extended research, training and information facilities. In the U.S.A., the Occupational Safety and Health Act of 1970 resulted in the creation of two new agencies, the Occupational Safety and Health Administration (OSHA) responsible for the inspectorate through ten regional offices and the National Institute of Occupational Safety and Health (NIOSH). Incidentally, the Robens Committee visited Sweden in 1971 and was much impressed by the range of research activities, from Chemistry to Psychology, in the National Institute of Occupational Health, and by the genuine efforts to organise interdisciplinary research.

The burgeoning of activity in occupational safety and health was one consequence of the steadily increasing affluence during the fifties and sixties. The level of prosperity in western countries became such that government, employers and trade-unions were able to turn their attention from the more strictly economic considerations of supplies of goods, profits and wages to wider issues of occupational health, welfare and quality of working life. During this period also there had been a rapid expansion of universities and there was an implicit need to find occupations for all these new graduates. Occupational Safety and Health is useful in this respect in that it can absorb graduates from medicine, engineering, natural sciences and social-sciences.

The degree of concern about occupational accidents and the available resources devoted to the matter have grown in parallel but one is not the cause of the other, both are consequences of a generally affluent society. Correspondingly, as economic growth diminishes as it has done over the past five years in all European countries, there is a corresponding need to question more closely the objectives, costs and benefits of accident research.

ACCIDENTS IN THE BROADER DEMOGRAPHIC CONTEXT

Figure 1 shows the incidence of death in relation to age as it was in 1900 compared with 1975. There has been a marked shift from death as something occurring steadily in all age groups to something which is very rare

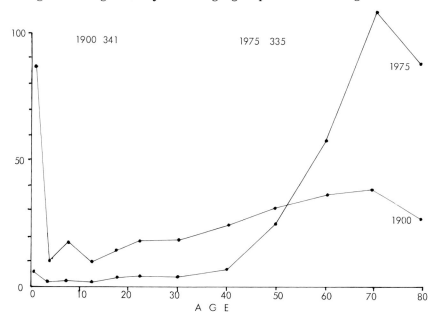

Fig. 1. Deaths in each year of age — males (CSO data) (thousands).

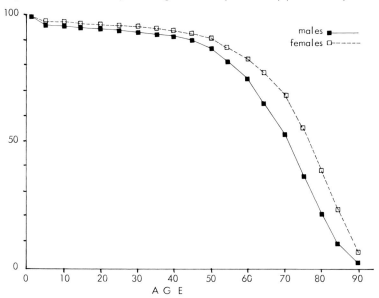

Fig. 2. U.K. life graph (CSO data) for the 1950 cohort.

under the age of 40 and not really common until after the age of 60. Figure 2 shows the same effect, if the present situation continues more than half of those born in 1950 will be still alive in 2020.

Consistently through the 1970's the proportion of total deaths due to violence in any form is less than four percent. Of the violent deaths less than three percent are due to occupational accidents, that is fatal occupational accidents account for just over one tenth of one percent of total deaths. Those of us engaged in occupational accident studies are aiming to reduce this figure. These data are for the U.K. but the situation is similar in all European countries. Figure 3 shows that fatal-accident rates are similar in Sweden and the U.K. and there is a downward trend, although only slight in

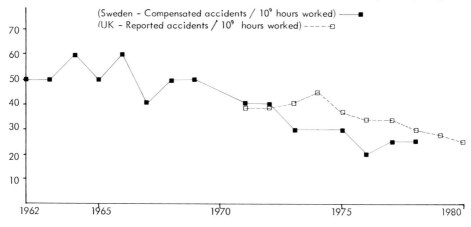

Fig. 3. Sweden (I.L.O. data). Fatal accidents in the manufacturing industry.

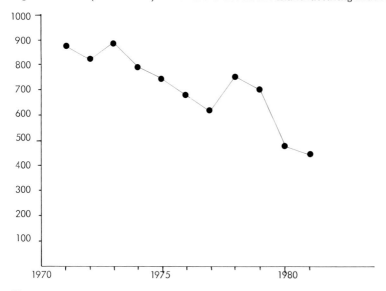

Fig. 4. U.K. (H.S.E. data). Employees killed in industrial accidents.

more recent years. There is another point of view, namely that proportions and percentages are not the issue but rather the number of individuals affected. Figure 4 shows that the number of employees killed in industrial accidents is still falling, but this is partly a reflection of reduced manufacturing activity. Figure 5 shows the situation in U.K. agriculture. Again the downward trend is levelling off at perhaps 70 deaths per year. More intensive study of the data of about 1970 led to the conclusion that unavoidable deaths are about 15 per year (Singleton, 1977). This category is defined as those events where neither improved equipment/job design nor improved training/awareness would have eliminated the accident. Usually the person involved took a reasonable risk but the unlikely event actually occurred as, of course, it is bound to on some occasions. The only way to reduce this kind of accident is to change the concept of a reasonable risk in the particular context. There is a limit to how far this process can go beyond which the situation becomes increasingly absurd as well as expensive.

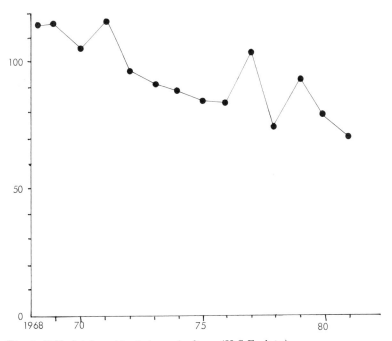

Fig. 5. U.K. fatal accidents in agriculture (H.S.E. data).

ACCIDENT RESEARCH IN SWEDEN

The advent of the Swedish Work Environment Fund and the reorganisation of the National Board of Occupational Safety and Health radically changed the research situation in Sweden. The budget of the Occupational Safety and Health Board increased from about 10 million S. Kr. in 1970 to 120 million in 1980. Correspondingly, the number of employees in the Labour

Inspectorate increased from 250 to more than 600 over this ten-year period. Accident research is, of course, only one of the topics of interest to the Fund and the Board. The Research and Development budget for 1981/1982 is shown in Fig. 6. The specific accident budget is only a small proportion of the total but of course, many other parts of the budget make their contribution to safety. (ASF, 1982). The striking aspects illustrated in this figure are the already-dominant chemical working environment problems and the very strong interest in psycho-social features, including work organisation and co-determination.

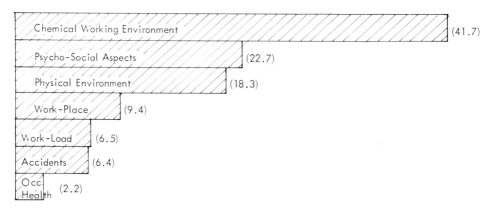

Fig. 6. Research and development budget (107.2 million S.Kr.): Swedish Work Environment Fund (1981/82).

In 1975, there was a seminar on occupational accident research in Saltsjobaden (A.S.F., 1976). The field was subdivided into the effects of legislation, behaviour, technology and organisational factors. Within this book Elizabeth Lagerlöf distinguished three approaches to accident research; behavioural models, epidemiological models and systems models. A year later, there was a French/Swedish symposium in Stockholm (ASF, 1977), in this book Dr Lagerlöf quotes a Fund Working Party: "there is a lack of theory in accident research and a subsequent lack of theory-testing research ... there is a need to develop conceptual models of accident causation". I am not convinced of the validity of the second part of this statement — a point which will be developed later in this paper. The first part however would be agreed by most workers in this field.

It is, of course, very difficult to distinguish cause and effect. The Fund policy makers seem to have started from the point of view that, since the scientific or academic side of accident research is not strong, it would be appropriate to ensure that studies stay close to reality so that findings are practically meaningful and readily applicable. This leads to broad cost effectiveness for research and development grants but it does not often lead to the generation of new theory. To make a practical contribution to a complex behavioural situation is difficult, to structure behavioural events into coherent theory is also difficult, to ask a research worker to do both simultanously is a challenge to which only the most exceptional individuals can rise.

There seems to have been some emphasis on the dangerous industries e.g. transport, forestry and explosives and there are a number of encouraging examples of positive effects on the real situation e.g. the reduction of 'kickback' accidents when using chain-saws. On the theoretical side there seems to have been little use of behavioural models, considerable development of expertise in epidemiological type studies and extensive interest in systems-type models. These latter seem to fit the accident situation well in that the multiple and serial nature of accident causation can be illustrated. However, in the pursuit of comprehensiveness, these models get more and more complicated and it becomes less and less feasible to make any kind of prediction from them.

THEORIES ABOUT ACCIDENTS

The theoretical position over the past fifty years has changed the emphasis from the kind of people who have accidents (accident proneness) to the kind of situation which results in an accident. This is partly due to the Swedish concentration on the environment as the locus of variables which might be changed to increase safety.

It is perhaps unfortunate that safety research has followed too closely the approved methodology of the physical sciences which is essentially a reductionist and iterative approach. The scientist conducts some observations, collects some data and generates a theory which explains how and why the data are related internally. The theory is then tested by predicting what will happen in specified controlled situations. This approach can never work in relation to safety. It is not possible, either ethically or conceptually, to devise controlled situations for which accident rates have been predicted and then check the predictions by noting the real events. The classical scientific method is inappropriate in its emphasis on the central role of prediction within validation. There can never be a theory which will predict an accident and even accident rates are subject to too many variables for prediction to be meaningful. Another basic assumption in classical science is that theories must be about causes. An event is seen as an effect for which there must have been a cause, if we can identify the cause then we can predict and control the event. Systems theories have at least demonstrated that this approach is not going to be fruitful in relation to accidents.

It does not follow that we must abandon hope of controlling accidents. The same problem occurs in other complex practical situations. The physician, for example, is often faced with a patient with a disease which he cannot readily identify, that is, the cause of the patient's symptoms is unknown. However, this does not mean that nothing can be done. The physician has certain general principles; the temperature must not be allowed to get too high, the body must not get dehydrated and so on. He can take action on the basis of these principles without waiting to identify the cause of the symptoms. Similarly in accident prevention we can take action to increase safety without waiting for a theory of accident causation.

Another user of this method is the process controller. In very complex multivariate control situations such as in nuclear power stations the failure of a particular set of sub-systems will lead to a pattern of function changes and alarm signals from which the sources of the trouble are not easily identified. Yet interference with the system on the basis of guesses about what has happened which later prove to be wrong may well make the situation worse. In these circumstances it is sometimes more effective to resort to control of key functions, viz. particular temperatures, pressures and voltages, and concentrate on keeping these within appropriate ranges. In any process control, the situation will be understood with hindsight because obviously nothing happens which is outside the laws of physics and chemistry. It is unusual and unpredicted sequences of events which lead to temporary failure of comprehension. The number of these possible, if unlikely, combinations is such that they cannot all be predicted, the controller must cope with the particular set which occurs.

By analogy, the safety specialist might be more effective if he were to concentrate rather less on looking for theories which will explain the cause of particular accidents and devote his attention to identification of the key parameters which can alter the probability of accidents generally. It can even be argued that this is the best that he will ever be able to do. The accident investigator should be a process controller rather than a research scientist in the traditional mould. This might be called a parameter control methodology rather than a cause identification methodology. In these circumstances

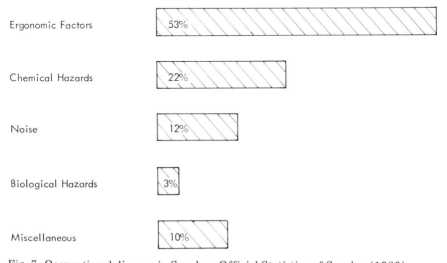

Fig. 7. Occupational diseases in Sweden: Official Statistics of Sweden (1980).

"conceptual models of accident causation" cease to be ends in themselves and become methods of identifying the relevant parameters and their appropriate ranges. In Sweden and in other countries there is already a move in this direction, for example in the development of the deviation concept in occupational accident control (Kjellén, 1983).

These comments refer only to accidents. There are other parts of the safety field where it is still appropriate to use the traditional scientific theories and methods, for example, in the understanding of toxic hazards (A.S.F., 1980). There is much concern about likely increases in occupational diseases due to increased use of chemicals although at present the major cause of such diseases is ergonomic factors, e.g., lifting and falling (Fig. 7).

ACCIDENT STATISTICS

There is currently much interest in the collation and distribution of accident data stimulated by the availability of computer systems connected into international networks e.g. "Euronet". If one tries to use data on an international scale it remains extremely difficult, if not impossible, to make any meaningful comparisons between countries. This is true even for the relatively unambiguous data about fatal accidents (Singleton et al., 1981). Unfortunately, it is much easier to set up new data-handling facilities than it is to reform and standardise ways of collecting data. The problem is not insurmountable but it is going to require extensive efforts on an international scale. The key to success is not to be too ambitious. To obtain standard data about fatal occupational accidents in a few selected industries across Europe would be useful progress over the next ten years. An enterprise with this limited objective would require extensive discussions about the procedures to be used and there would need to be standardised training of the investigators who study the particular events. There would also need to be extensive liaison and agreement between governments, employers, trade-unions, police and legal systems.

CONCLUSION

For the remainder of this century occupational accident research is likely to proceed on about the same scale as at present. Accident rates will probably continue to decrease slowly mainly because of advancing technology but with some influence from the behavioural measures of better ergonomics and better training. There may well be a considerable increase in investment in safety generally but this is likely to be devoted to chemical hazards and to "non-professional" accidents rather than to accidents at work. Occupational accident studies will be much more tightly controlled in cost-benefit terms. There should be considerable advance in the sophistication of epidemiological studies. Some limited progress should be made in the international standardisation of data.

Theoretical studies of the systems type will be orientated more towards methods of accident control rather than methods of explaining particular accidents. Behavioural type theory may benefit from studies of wider societal issues such as crime and alcoholism, but here also new insights will lead to better control rather than more precise prediction.

This may not appear to be an enormously exciting picture of the future but it is a very interesting one with profound intellectual challenge and the rewards of real benefits to the community.

REFERENCES

A.S.F., 1976. Occupational Accident Research. Swedish Work Environment Fund, Stockholm.
A.S.F., 1977. Research on Occupational Accident. Swedish Work Environment Fund, Stockholm.
A.S.F., 1980. Solvents in the Work Environment. Swedish Work Environment Fund, Stockholm.
A.S.F., 1982. Programme of Activities and Budget, 1981—1984. Swedish Work Environment Fund, Stockholm.
Cresswell, C.W. and Froggatt, P., 1963. The Causation of Bus Driver Accidents. Oxford University Press, Oxford.
Greenwood, M. and Woods, H.M., 1919. The incidence of industrial accidents upon individuals with special reference to Multiple Accidents. Industrial Fatigue Research Board Report No. 4. H.M.S.O., London.
Kjellén, U., 1983. Analysis and development of corporate practices for accidents control. Stockholm Royal Institute of Technology: Report No. Trita - AVE 0001.
Singleton, W.T., 1977. Accidents in agriculture. British Association Symposium on Human Error. Aston University: A.P.D. Note 62.
Singleton, W.T., Hicks, C. and Hirsch, A., 1981. Safety in agriculture and related industries. Aston University: A.P.D. report 106.

ABSTRACT

Setting Priorities for Occupational Health and Safety Research

LYNN STEWART HEWITT and JUDITH E. MAKI EVANS

Alberta Occupational Health and Safety Division, 10709 Jasper Ave., Edmonton, Alberta T5J 3N3 (Canada)

This organization recently completed a study to identify research areas with the highest potential significance for occupational health and safety in Alberta. Based on interviews and questionnaires, 65 occupational health and safety professionals from government, industry and universities judged the potential value of undertaking research in each of 58 areas.

A high level of consensus emerged on the relative importance of these research issues. In general, research related to the *evaluation of preventive programs and strategies* received the highest ratings of potential value while research into *physical and biological hazards* received the lowest ratings. Participants emphasized the need for research into practical problems and generally rejected further study of well-researched issues.

EVALUATION OF ACCIDENT PREVENTION STRATEGIES: METHODOLOGY AND PRACTICAL RESULTS

IS SAFETY TRAINING WORTHWHILE?

A.R. HALE

University of Aston, Dept. of Environmental Occupational Health & Safety, Gosta Green, Birmingham B4 7ET (United Kingdom)

ABSTRACT

Hale, A.R., 1984. Is safety training worthwhile? *Journal of Occupational Accidents*, 6: 17—33.

Training has become almost an axiomatic part of accident prevention strategies, but it is questionable whether the level of analysis and the development of training programmes is sufficiently high to produce the return expected. Without this investment safety training is in danger of getting a bad name. The paper reviews the safety training provided in the UK and illustrates with examples of research results the mismatch between the requirements of effective training and the provision given.

INTRODUCTION

The U.K. Health and Safety at Work (etc.) Act, 1974, states as one of its general duties that the employer must ensure "the provision of such information, instruction, training, and supervision as is necessary to ensure so far as is reasonably practicable, the health and safety at work of his employees" (Section 2 (2c)). In addition to this sweeping general provision, Acts governing health and safety at work, and the numerous regulations made under them, have frequently made use of the concept of the appointment of a "competent" or "authorised" person to carry out particular tasks or supervise the work of others in the name of safety. Such legislative provisions have a long history. The first use of the term in relation to safety in the 19th Century was in relation to steam boilers which were required to be inspected by competent persons at regular intervals. Thereafter the strategy became increasingly used in relation to dangerous machines and later for work with dangerous chemicals and radiations and in hazardous environments. At the present day, almost 200 sections of Acts and regulations specify their employment in the various sectors from agriculture and mining, through construction, dock work and offshore work to factories and offices. Table 1 shows the range of activities covered.

Even so the use of the strategy is very patchy with considerable detail being given in some areas and nothing in others which are arguably equally dangerous. For example, setting of power presses is governed by a detailed

TABLE 1

Safety activities requiring competent or authorised persons by law

Activity	Examples
Management	Mines and quarries
Safety supervision	Shipbuilding, construction
Medical supervision	Radiation, lead workers
Examination and testing	Lifting machinery, pressure vessels
Skilled judgement or measurement	Opening wool bales, radiation monitor
Inspection	Government inspectors
Machine setting	Power presses, abrasive wheels
Machine operator	Cranes, mine hoists, air locks
Operation in dangerous environments	Demolition, mine rescue, diving

set of regulations which include a training schedule while injection moulding machines have no specific provisions. Regulations say nothing about needing to be competent to drive fork lift trucks or farm tractors, but do for coal mine face conveyor drivers.

Considering the widespread use of the "competent person" strategy it is surprising that the word has not received a satisfactory and detailed definition in English law*.

Two definitions of "competent" have been given in decided cases:

(1) "A man who on a fair assessment of the requirement of the task, the factors involved, the problems to be studied and the degree of risk of danger implicit, can fairly, as well as reasonably, be regarded by the manager, as competent to perform such an inspection". (Brasier v Skipton Rock Company, 1962 ALL ER/955 concerning the inspection of quarries).

(2) "I think that a competent person for this task is a person who is a practical and reasonable man, who knows what to look for and knows how to recognise it if he sees it". (Gibson v Skibs A/S Marina 1966. ALL ER 476, concerning the inspection of lifting gear).

Both of these definitions given by judges are remarkably circular and it is interesting that neither of them mentions training as a requirement for competence. Of the 200 or so regulations and sections which specify the employment of a competent or authorised person, only some 30 give any details of the sort of training which should be given in order to achieve that state and it is only in the Regulations produced since 1965 that the details become sufficiently explicit to be useful in planning training. Even here the wording is usually very general (e.g. to be aware of the risks, and precautions to be observed; to understand the safety devices and their operations, etc.). Alternatively, the law passes the buck by calling for a specific qualification (e.g.

---

*While some words used in statutory provisions in the U.K. are defined within those provisions, the majority are left to the definition of judges in particular cases who operate on a system of precedent. "Competence" is normally left to the latter.

mine managers, divers, first aiders) leaving the organisation giving it to decide what the training shall be.

The Committee on Safety and Health at Work chaired by Lord Robens which reported in 1972, and upon which current British legislation is based, stated in their report: "Most people are agreed that safety training is of vital importance. There is no unanimity about what in practice should follow from that proposition".

A review of the literature of health and safety reveals that this statement is still largely true. No text on safety is without the exhortation to carry out training in order to improve accident prevention, but the subjects which they then go on to talk about are many and varied and there is no agreement about which should be included under the heading of 'safety training' and which should be separated out under a different heading, such as 'motivation'. Few talk at all about proof of its effectiveness. It would seem that the subject of safety training is one which goes alongside motherhood and progress as self-evidently a good thing. It is a basic tenet of faith of the safety movement that safety training must take place, and yet general reviews of the subject (e.g. Hale and Hale, 1972; Cohen et al., 1979; Surry, 1969; Heath, 1981; Jonah et al., 1982; Edwards and Ellis, 1976) all start off their discussion with the comment that remarkably few studies have been published which evaluate the effectiveness of safety training programmes.

This paper examines the confusion over the meaning of safety training, reviews the relevant evaluation studies, and proposes a classification which, it is hoped, will overcome some of these confusions and suggest further studies. The classification is based on an information processing model of behaviour in the face of danger and hence does not emphasise the motivational aspects of the subject as much as the cognitive ones (but see the section Decision to Act, below). In this context training is considered as any activity which aims to increase the *capacity* of a person to respond in ways appropriate to the situation facing them. This includes the capacity to respond more rapidly and the capacity to evaluate situations more effectively, as well as the capacity to respond in new ways. Motivational factors influencing *choice* between known and available actions and influencing desire to respond are therefore formally excluded, although in practice it is often difficult to know whether a change in behaviour has resulted primarily from new knowledge about actions and their consequences or from a change in the attractiveness of such an action. The two are also clearly interrelated (e.g. Fishbein and Ajzen, 1975).

The paper does not review the few studies evaluating training of those whose job is to advise on safety or to negotiate over it, e.g. safety practitioners, inspectors of factories, safety representatives, etc. (see Kamienski, 1978; Cook, 1980). It concentrates on the training of those whose actions directly affect their own safety or that of others for whom they are responsible. As such the potential trainees are workers, process controllers, supervisors, managers and designers (although no studies were located concerned with the last group).

The paper does not review the methodological problems of carrying out evaluation research, although the paucity of reported research studies is perhaps witness to the complexity of those problems.

AGE, EXPERIENCE, TRAINING AND ACCIDENTS

From the earliest studies of personal factors and accident experience (e.g. Newbold, 1926) it has been apparent that young people and inexperienced people are more likely to suffer accidents than older and more experienced workers. Studies have generally shown that experience on the specific job is the more important variable, but that age has an independent effect in the years up to the mid twenties, perhaps indicating the influence of some broader experience of work and life as well as a greater maturity of approach to it (e.g. Brown and Ghiselli, 1948; Adelstein, 1952; Van Zelst, 1954; Powell et al., 1971). The study by Powell et al. (1971) indicated that frequency of repetition of tasks, and time on task were also related to accident liability. These results indicate that there is a learning process which naturally goes on as people gain experience of a task, and that that increase of experience produces a reduction in accidents. The potential is clearly there for systematising that process through training, and hence achieving the reduction in accident rate faster. Van Zelst's study (1954) is one which demonstrates this effect in comparing a trained group of workers with an untrained control group.

GENERAL JOB TRAINING

If we turn to the literature on general job training, and examine the studies which have reported the effect of training upon accident rate as well as upon task performance, the picture is not altogether clear. Hale and Hale (1972), reviewing the industrial accident research literature, indicated that there were almost as many studies showing that job training had no effect on safety as studies showing a positive effect. Indeed, there were occasions when those who had received training had higher accident rates (e.g. Cheradame, 1967; Rotta et al., 1957). Studies in road safety (e.g. Raymond and Tatum, 1977; Jonah et al., 1982) lead to the same conclusion. However, it is clear that much of this confusion is due to the variation in the quality of the training given, rather than to any intrinsic lack of relationships between training and job safety. The training programmes which were conducted with due regard to the principles of good training design (for example Shaw and Sichel, 1971; Van Zelst, 1954) showed significant reductions in accidents in trained groups as opposed to untrained. In a study comparing matched pairs of high and low accident rate companies, Cohen et al. (1975) found that the low accident group had more, and more varied, training and safety training programmes.

SAFETY TRAINING

The studies cited above all treated training as a unitary variable and did not set out specifically to design the training to reduce accidents. They simply evaluated the effect of a training programme which was already going on. If we turn to studies which specifically concern safety training as opposed to general job training, the confusion and lack of definition mentioned above becomes even more marked. Discussions of accident case histories, showing of safety films, demonstration of safe methods of working, courses of

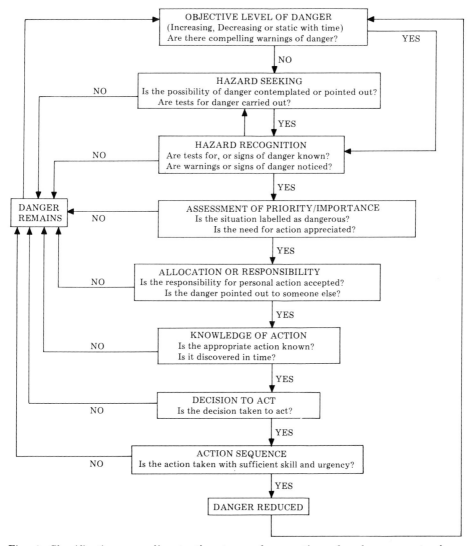

Fig. 1. Classification according to the stages of perception of and response to danger (adapted from Hale and Perusse, 1977).

defensive driving, and programmes of gymnastics aimed at improving reaction time are all labelled "safety training". Specific objectives are rarely mentioned, but it is clear that such a range of training cannot all be aimed at the same behaviour change.

Before proceeding to review the studies it is useful to present a framework to classify them and to make sense of the findings. Figure 1, adapted from Hale and Perusse (1977), provides a classification according to the stages of perception of and response to danger, which will be used as the framework for this paper. The model suggests seven areas of behaviour which are open to change and improvement through some sort of behaviour modification. The sixth division (Decision to Act) is the one which falls most into the area of motivation as opposed to training, but there are clearly interactions between the two in most areas. This paper will concentrate on the safety training aspects and mention safety motivation only in passing and where it is inextricably bound up with training. This division is far from clear-cut in research studies, and is arguably one which should not be made rigorously in practical attempts to alter behaviour since the two areas interact and can reinforce each other.

1. Hazard seeking

Where danger is not "obvious" an individual may have to go through an active search process to find if it is present. Studies by Perusse (1978) show that this active searching and questioning process is governed by motivational factors which include the attraction of competing activities, interest and the perceived subjective control which individuals feel they have over the task they are engaged in. A study by Duncan and Gray (1975) in the chemical industry shows that operators can be trained to carry out more systematic error seeking, in their case the verification of information presented by, or inferred from, process control panel displays. This is equivalent to training people to mistrust the situation in which they find themselves and to spend more time seeking out hazards in it than they did before. An analogous situation is the teaching of defensive driving which has been shown to have some effect on road accidents, (e.g. Payne and Barmack, 1963). However, this also incorporates training under later headings. Training in where to look for hazards, e.g. before starting up a machine, would also come under this heading.

2. Hazard recognition

Hazard recognition is very closely related to hazard seeking. The decision to seek out hazards may be based upon a preliminary decision that "something is wrong". The final decision that danger is present is then based upon putting together the evidence of one or more signs or symptoms revealed during that search.

Behind hazard recognition, therefore, there is a requirement to understand the cause and effect network which leads from those signs and symptoms to the occurrence of the potential disease, accident or disaster. The skills of searching and inspection therefore need to be combined with the cognitive, diagnostic function of putting those indicators together and making sense of them. Studies by Blignault (1979a, b) in the South African mines showed that the ability of miners to detect potential rock falls was an important factor in accident causation and a factor which seemed to differentiate between novice and experienced miners. He found that it was possible to improve the ability of novice miners to detect potential falls by training in formal search strategies in the laboratory, and also by practice on slides of actual mining situations. Improvement was significant only in the group given practice, not in one just informed of what to look for.

Embrey (1979), reviewing the evaluation studies on training for industrial inspection tasks, underlined the value of cueing and of feedback of knowledge of results in the training of inspectors. It would seem likely that training in hazard recognition would also benefit from similar training strategies.

However, recognition is not purely a matter of visual search. Pilot studies I have made show that novice inspectors treat it as such and hence only spot a small proportion of hazards in work places (see also Hale, 1978). Perusse (Hale and Perusse, 1978) studied how people go about spotting hazards. He asked people to verbalise the process they went through in searching a picture for hazards. It was clear that the process was one of hypothesis generation and testing and that there were individual differences in strategy used. Some set up a hypothesis, e.g. "there could be children around; what might they do?" and then searched the picture for appropriate clues. Others looked at themes, e.g. electrical dangers, tidyness, design of equipment. While others seemed to use a global 'Gestalt'; "that looks nasty to me"; and only specified what the hazard was when questioned. Clearly there is scope for research into which strategy is the best and then for training in it.

Evaluation studies of training of process control operators (Shepherd et al., 1977; Marshall et al., 1981) have shown the importance of training in the heuristics of problem solving and fault diagnosis. Their work showed that operators trained in these techniques were far better at the diagnosis of faults that they had not seen before than operators who had merely received a general training in the theory and practice of the plant operation. They point out that operators who have to start from theory must postulate a fault and then deduce the symptoms, whereas reality presents them with the symptoms from which they must infer the fault.

A number of evaluation studies of training programmes can be classified, from a reading of the training contents, under the heading of hazard recognition and cause/effect training. For example Zohar et al. (1980) provided feedback from audiometric testing to workers in noisy workshops in an attempt to increase the wearing of earplugs. It is apparent that the evidence of temporary threshold shift acted as a reminder of the cause/effect link, and that, together with the change in priority induced, produced the dramatic increase in wearing noted in the study. A similar result was obtained by Merriman (1977). Lagerlöf (1982) used the reporting of near-accidents and the feedback of the information to work groups in a similar way and found not only a higher risk-consciousness, but a reduction of accidents in those groups compared with control groups which did not have the same reporting system. (She reported that this effect was produced without the work force asking for or achieving changes in the work methods or environment). The studies by Komaki et al. (1978), Rubinsky and Smith (1973) and Edwards and Ellis (1976) all indicate that the main content of the safety training they provided was a discussion or demonstration of the link between hazard and accident. Rubinsky and Smith (1973) simulated this accident on a modified bench grinder by spraying people with water if they stood in front of the grinder at start-up, rather than with the disintegrating grinding wheel of the real accident. Komaki et al. (1978) provided feedback of observed safe and unsafe behaviour over subsequent periods as well as the initial training and it is clear from a later study (Komaki et al., 1980) that the feedback was the more important element. All three studies showed an improvement in their criterion measures though, in the study by Edwards and Ellis (1976) on road safety, the improvement was in traffic violations rather than in accident experience for most groups.

3. Assessment of priority/importance

Assessment of priority and importance of the danger is an area of great overlap between training and motivation. Perusse (1980) showed that individuals' assessments of hazards depended upon two major factors which he called "Scope for Human Intervention" and "Dangerousness". The former usually dominated, although for some individuals and for some hazards the latter swamped the picture (cf. Fishhoff et al., 1978, and the position of nuclear power in people's assessment "space"). It is clear from his study that cognitive and affective attributes are woven together in both factors, and that a concentration solely on trying to convince people that the consequences and likelihood of the harm were greater than they had thought would not necessarily have the desired effect.

Pirani and Reynolds (1976) in a study of various methods of altering behaviour towards protective clothing showed that, although the motivational changes such as fear-arousing films and discipline all had quite marked initial effects on the wearing of protective equipment, it was the participation in role-playing of various scenarios, including discussion on compensation, which had the longest lasting effect. Pirani and Reynolds state that the workforce knew of the hazards and the precautions before the interventions. Therefore the effect of the role-playing must largely have been in producing a better understanding of the importance of the consequences of the harm and of the way it was controlled by the protective equipment. This superiority of knowledge-based change over fear-based is in line with the extensive work on health education and attitude change (e.g. Levanthal, 1970; Rogers and Mewborn, 1976; Sell, 1970) which shows that fear arousal may not be very effective especially for potential victims as opposed to bystanders and may even be counter productive if it floods the individuals' cognitive faculties to such an extent that they do not take in the information about how to act safely (Wilkins and Sheppard, 1970). The work of Fischhoff et al. (e.g. Lichtenstein et al., 1978; Slovic et al., 1980) and that of Tversky and Kahneman (1972) has shown the effect of 'recency' and 'availability' on people's judgement of hazard. These concepts are descriptions of the changes in priority brought about by the experience of hazards and accidents, either personally or vicariously. Training therefore seems to need to provide the mental image of an accident which can be triggered to stimulate a high rating of priority, but that image must not be too horrific or it will not be either accepted or released from suppression. The results of McKenna's evaluation of first-aid training indicate that this may be one way of providing such a balance which influences accident incidence (McKenna and Hale, 1982).

4. Allocation of responsibility

Allocation of responsibility covers the correct acceptance of the responsibility for action by an individual, or its correct allocation to someone else through a suitable communication or warning. The importance of this area is illustrated by the research of Abeytunga (Abeytunga and Hale, 1982) into the training needs of construction site supervisors in health and safety. In a series of questions supervisors were asked about the hazards or "accident symptoms" present on their site during a joint inspection with the researcher. Supervisors were assessed to see if they recognised accident symptoms and, if not, whether they accepted them as hazards when they were pointed out; whether they accepted responsibility for taking any action (including warning or instructing others to remove the hazards), and if not who else they thought should be responsible; whether they knew the action to remove the hazard, and if not whether they agreed with the proposed action; and finally why they thought the hazard was still present. The most striking finding is that, for 64% of the hazards present, action was regarded by the supervisors as being the responsibility of someone else. This was despite the fact that the supervisors were, according to their companies, responsible for everything which happened on the sites which were being inspected. The clash of views is striking. It is a further question as to which view is correct, but it is clear that any training aimed at site supervisors which did not recognise this problem would be bound to be unsuccessful.

Very few of the published evaluation studies seem to address training in this area. It is implicit in the frequent specification of training in legal and policy responsibilities as a necessary component of safety training courses, but the only evaluation studies which specifically mention it are those by Lagerlöf (1982) who reported a positive change in workers' acceptance of responsibility and feeling of control over their health and safety following the programme of near-accident reporting and feedback which she studied, and the work by McKenna — McKenna (1978) and McKenna and Hale (1982) — showed that the main effect of first aid training on attitudes was in the area of a more rational allocation of responsibility for accident causation and prevention. The finding of Kjellén and Baneryd (1983) that workers involved in their discussion groups had a more positive attitude to safety measures could be interpreted as a change in this area.

5. Knowledge of action

Knowledge of what to do to avoid or remove the danger is one of the most obvious areas for training and hence it is covered implicitly in a number of the evaluation studies reported. Conard et al. (reported in Cohen et al., 1979) informed workers using styrene of the correct working procedures and the reasons for them, and reported very significant increases in the use of those methods. They indicated that this was more in the nature of prompting or refresher training than of initial training.

The work of Leventhal et al. (1965) has indicated that knowledge of the course of action to be followed is a more important ingredient in health education than the arousal of fear. In the same area, Rogers and Mewborn (1976) have demonstrated the effectiveness of convincing people of the efficacy of control measures in getting them to change their behaviour.

6. Decision to act

The decision to act is an area of motivation rather than of training, since it is a matter of weighing the competing courses of action which have been thought of or presented. There is ample evidence from the work of the behaviour modification school (e.g. McKelvey et al., 1973; Komaki et al., 1978; Sulzer-Azaroff et al., 1980; Zohar and Fussfeld, 1981, etc.) that carefully organised programmes of feedback of information and the use of positive incentives such as tokens or praise can produce a significant increase in the desired behaviours and a significant reduction in accidents. As indicated above, the negative incentives of fear and discipline are less effective, particularly if their use is sustained over long periods (e.g. Pirani and Reynolds, 1976; Tourigny, 1980).

7. Action sequence

The final area, action sequences, covers the skills necessary to carry out the safe behaviour and their co-ordination. They are a matter of concern particularly where complex psycho-motor skills must be used at infrequent intervals in emergencies. Examples are emergency shut-down procedures in plants, disaster planning and evacuation, first-aid treatment, and rescue. The effectiveness of training in inculcating some of these skills, and the need for overlearning and refresher training in achieving retention of them, is well documented (e.g. Stammers, 1979; Weaver et al., 1979; Winchall and Safer, 1966). In some of the areas, notably in training for disaster plans and evacuation, the evidence on effectiveness of training does not appear to exceed the anecdotal level. Also, most industries in which breakdown of such skills can be catastrophic, e.g. air transport and nuclear power, recognise the limitation of training by their introduction of procedural checklists and automated sequences.

A CASE STUDY: MANUAL HANDLING TRAINING

A recent study carried out at Aston to evaluate manual handling training (Mason, 1982) illustrates the interaction between the different areas of safety training mentioned above, and reveals some of the confusion and waste produced by poor analysis of the training needs and problems.

A review of the practice of manual handling training in British industry, and of the literature relating to it, both official and popular, revealed that the prevailing image of the problem was that manual handling produced back injury and that the solution lay in the adoption of a drill of movement, frequently known as the "six point drill". (See Table 2). In other words, the problem has been placed into the seventh category, namely a problem of the correct action sequence, which was designed to protect the lumbar spine from excessive loading. Typical training consisted either of handing workers cards with the drill written on them or showing brief films, or demonstrations, the whole normally occupying no more than an hour or two.

Comparison of this stereotyped view with the reality of injury statistics and with the complexity of body movement required to handle loads safely, showed that the range of possible injuries was not just to the back, but to all parts of the surface as well as the muscles and joints of the body. It also revealed the importance of the problem of cumulative strain from bad posture and movement and its long term effect on arthritic and rheumatic

TABLE 2

Six point drill

(1) Put one foot alongside the object and one behind.
(2) Keep the back straight.
(3) Tuck in your chin so the head and neck continue the straight back line.
(4) Get a firm grip with the palms of your hands.
(5) Draw the object close to you, arms and elbows tucked into centre body weight.
(6) Lift straight with a thrust of the rear foot.

National Safety Council (1976).

complaints. Laboratory studies also showed that the conventional training produced movements which were not even correct by the criteria of the six point drill itself, let alone by the criteria derived from "kinetic handling" theory. A study using pictorial presentation of postures to look at subjects' concepts of good and bad posture showed that there were characteristic misconceptions. Notably a significant number of subjects thought that good posture was equated with positions in which the arms rather than the back were doing the work. Similarly there was a confusion of relaxed posture with sagging or slumped postures. Also some people thought that postures which showed a lot of brute force being applied were good.

Examination of the language used for describing good posture during training and in instruction manuals also revealed confusions. One term which is widely used is "keep a straight back". The study showed that different interpretations were given to this instruction, revealing confusion between erectness and lack of curvature of the spine. Other parts of the study looked at the posture adopted by subjects when they were asked to copy movements demonstrated, or carry out instructions in manuals. It was very apparent that there was a gap between people's ability to describe a posture and to adopt it. People were very surprised when shown videotapes of their movements and thought that they had adopted completely different postures to the ones shown.

The main training evaluation study looked at the effect of a one week training course in handling methods provided for future kinetics instructors by the Royal Society for the Prevention of Accidents. To assess it, specific criteria were defined for good and bad movements, and the subjects were videotaped before the training and at the end of the course. Their movements were analysed using the kinetic criteria to assess the degree of change. All training was carried out by one trainer, and the subjects were an unselected population of industrial supervisors and managers sent by their companies on the training course.

TABLE 3

60 subjects tested on 11 criteria of good moment, $N = 658$* (Mason, 1982)

Before Training		After Training	
Failed	493	Passed	341
		Improved	21
		Failed	131
Passed	165	Passed	156
		Failed	9
Pass rate	25.1%	78.7% (Passed or improved)	

*Results for one subject on two criteria were not determinable from the video.

The overall results are contained in Table 3. This indicates that there was a significant improvement in the trainees over the period. However, only 10% were passing on all criteria by the end of the one week of training, and there was still a failure rate of 28% and 43% on the two most important criteria. It was not possible to carry out a retest at a subsequent date, and hence this result perhaps presents an optimistic picture of the effectiveness of the training.

The overall conclusion of the research was that much training had tackled the wrong problem; that recognition of good and bad posture and body awareness were far more important than any attempt to learn an action drill. A pilot study using a video feedback during the one week training indicated that there was considerable potential for training improvement when these particular problems were tackled directly. Intensive training over a week including extensive practice did change movement patterns, but not totally and hence one or two hour courses would be unlikely to get very far.

METHODS OF TRAINING

A few of the safety training evaluation studies have specifically aimed to compare the effectiveness of different training methods in producing an outcome, either in changed behaviour in the face of danger, or in a changed accident rate. For example Rubinsky and Smith (1973) found that experience of their simulated accident was a better training aid than demonstration of it or a description of it, when the criterion of number of "accidents" and retention over time were used. Leslie and Adams (1973) found that direct face-to-face training and demonstration was better than tape/slide or video presentation in avoiding accidents on a punch press simulator. Glassman (1965) found a teaching machine better than face-to-face training for road hazard training. Mason (1982) found video feedback of manual handling a better training method than simple practice. Stammers (1979) discusses the provision of simulators for power plant operators, and questions the need for very high fidelity simulators in cost/benefit terms. Iacono (1967) found group discussions led by psychologists better at changing attitudes to safety clothing than those unled or led by a non-psychologist. Fugal (1950) showed individual training to be better than group methods in a range of industrial tasks. This catalogue of unconnected results is hard to interpret. The apparent contradictions are doubtless partly because the areas being trained differed and partly because the quality of the training varied as well as its format and methods. All that can be concluded is that far more studies are necessary before anything can be concluded.

DISCUSSIONS AND CONCLUSIONS

The low number of studies located and discussed in this paper indicates the dearth of research in the evaluation of safety training in industry. Undoubtedly one can draw some inferences from the rest of training research but, until the analysis of safety training needs is carried out more systematically, using such frameworks as I have presented, it is not always easy to see which bits are relevant. Clearly part of the problem is the difficulty of carrying out evaluation research in safety. The use of criteria such as accident rates is fraught with problems because of the rarity of accidents and hence the need to use long follow-up periods, which in turn introduce the difficulty of eliminating other changes over the research period which could contaminate the evaluation. But other criteria are possible, such as changes in specific behaviour or attitudes and it is distressing that so little research using such variables has been published. The works reviewed above do not provide the solid framework and foundation upon which the current vast edifice of safety training can be confidently built. No general conclusions can be drawn about the relative importance of training in the different areas in different situations, nor about the effectiveness of different means of training, nor between competing explanations of the effects produced. Conclusions must largely be limited to the specific cases studied.

It is disappointing to have to conclude a review such as this so negatively. The potential for training is clearly perceived by all involved in safety, but it would appear that this "obvious" potential has blinded those working in the area to the need for the careful evaluation which would allow the potential to become fully realised. The overwhelming need is for field-based studies of well-designed training programmes aimed at specific objectives relevant to health and safety.

REFERENCES

Abeytunga, P.K. and Hale, A.R., 1982. Supervisors' perceptions of hazards on construction sites. 20th Annual Congress of the International Association of Applied Psychology, Edinburgh.

Adelstein, A.M., 1952. Accident proneness: a criticism of the concept based upon an analysis of Shunter's accidents. J. R. Stat. Soc. 115 (3): 354—410.

Blignault, C.J.H., 1979. The perception of hazard: 1. Hazard analysis and the contribution of visual search to hazard perception. Ergonomics 22 (9): 991—999.

Blignault, C.J.H., 1979. Hazard perception: 2. The contribution of signal detection to hazard perception. Ergonomics 22 (11): 1177—1183.

Brown, C.W. and Ghiselli, E.E., 1948. Accident proneness among streetcar motormen and motor coach operators. J. Appl. Psychology, Vol. 32: 20.

Cheradame, M.R., 1967. Incidence of selection and training of personnel in relation to prevention of accidents at work. European Coal and Steel Industry, Luxembourg.

Cohen, A., Smith, M.J. and Anger, W.K., 1979. Self-protective measures against workplace hazards. J. Safety Research 11 (3): 121—131.

Cohen, A., Smith, H.J. and Cohen, H.H., 1975. Safety program practices in high v. low accident rate companies. An interim report. Health Education and Welfare Publication NIOSH 75-185.

Cook, D., 1980. A pilot study to investigate the effect of the TUC safety representatives' training course on the safety representatives' perceptions of their role. MSc dissertation. Department of Environmental and Occupational Health, University of Aston, Birmingham.

Duncan, K.D. and Gray, M.J., 1975. An evaluation of a fault finding training course for refinery process operators. J. Occup. Psychology 48: 199—218.

Edwards, M.L. and Ellis, N.C., 1976. An evaluation of the Texas driver improvement training program. Human Factors 18 (4): 327—334.

Embrey, D.E., 1979. Approaches to training for industrial inspection. Applied Ergonomics 10 (3): 139—144.

Fishbein, M. and Ajzen, I., 1975. Belief, Attitude, Intention and Behaviour. Addison-Wesley Publishing Co., Reading, Mass.

Fischhoff, B., Slovic, P., Lichtenstein, S., Read, S. and Combs, B., 1978. How safe is safe enough? A psychometric study of attitudes towards technological risks and benefits. Policy Sciences 9: 127—152.

Fugal, S.R., 1950. Relationship of safety education to industrial accidents. PhD Thesis. Yale University, USA.

Glassman, J., 1965. The effectiveness of a teaching machine program as compared with traditional instruction in the learning of correct responses to hazardous driving situations. PhD Dissertation, New York University.

Glendon, A.I. and McKenna, S.P., 1979. First Aid Community Training (FACT) (U.K.). Final report to St John Ambulance. Department of Safety and Hygiene, University of Aston, Birmingham.

G.B.: Department of Employment, 1972. Safety and health at work: report of the committee 1970—72. Chairman Lord Robens. Cmnd 5034, H.M.S.O., London.

Hale, A.R., 1978. Collecting information. In: Working for Safety. BBC Publications, London.

Hale, A.R. and Hale, M., 1972. Review of the industrial accident research literature. Research Paper. Committee on Safety and Health at Work. HMSO, London.

Hale, A.R. and Perusse, M., 1977. Attitudes to safety: facts and assumptions. In: Phillips, J. (Ed.), Safety at Work, SSRC, Conference Paper No 1. Centre of Socio-legal Studies, Wolfson College, Oxford.

Heath, E.D., 1981. Worker training and education in occupational safety and health: a report on practice in 6 industrialised western nations. US Department of Labour Occupational Safety and Health Administration.

Iacono, G., 1966. The dynamics of resistance of individuals in groups to the utilisation of the means of protection. In: Human Factors and Safety In Mines and Steel Works. Studies in Physiology and Psychology of Work No. 2. European Coal and Steel Community, Luxembourg.

Jonah, B.A., Dawson, N.E. and Bragg, B.W.E., 1982. Are formally trained motorcyclists safer? Accident Analysis and Prevention 14 (4): 247—255.

Kamienski, A., 1978. Training in improved working environments: progress and problems — a summary and evaluation. Swedish Working Environment Fund.

Kjellén, U. and Baneryd, K., 1982. Changing the local safety and health practices at work within the explosives industry. Ergonomics 26 (9): 863—877.

Komaki, J., Barwick, K.D. and Scott, L.R., 1978. A behavioural approach to occupational safety: pin-pointing and re-inforcing safety performance in a food manufacturing plant. J. Applied Psychology 63 (4): 434—445.

Komaki, J., Heinzmann, A.T. and Lawson, L., 1980. Effect of training and feedback: component analysis of a behavioural safety program. J. Applied Psychology 65 (3): 261—270.

Lagerlöf, E., 1982. Accident reduction in forestry through risk identification, risk consciousness and work organisation change. 20th Congress of Applied Psychology, Edinburgh.

Leslie, J. and Adams, S.K., 1973. Programmed safety through programmed learning. Human Factors 15: 223—236.

Leventhal, H., 1970. Findings and theory in the study of fear communications. In: Berkowitz (Ed.), Advances in Experimental Social Psychology, Vol. 5, New York, Academic Press.

Leventhal, M., Singer, R. and Jones, S., 1965. Effects of fear and specificity of recommendations upon attitudes and behaviour. J. Personality and Social Psychology 2: 20—29.

Lichtenstein, S., Slovic, P., Fischoff, B., Layman, M. and Combs, B., 1978. Judged frequency of lethal events. J. Experimental Psychology: Human Learning and Memory 4 (6): 551—578.

McKelvey, R.K., Engen, T. and Peck, M.B., 1973. Performance efficiency and injury avoidance as a function of positive and negative incentives. J. Safety Research 5 (2): 90—96.

McKenna, S.P., 1978. The effects of first aid training on safety: a field study of approaches and methods. PhD Thesis, University of Aston, Birmingham.

McKenna, S.P. and Glendon, A.I., 1982. The benefits of first aid training for road safety. Department of Occupational Health and Safety, University of Aston, Birmingham.

McKenna, S.P. and Hale, A.R., 1982. Changing behaviour towards danger: the effect of first aid training. J. Occupational Accidents 4 (1): 47—60.

McKenna, S.P. and Hale, A.R., 1981. The effect of emergency first aid training on the incidence of accidents in factories. J. Occupational Accidents 3 (2): 101—114.

Marshall, E.C., Scanlon, K.E., Shepherd, A. and Duncan, K.D., 1981. Panel diagnosis training for major hazard continuous-process installations. The Chemical Engineer: 66—69.

Mason, I.D., 1982. An evaluation of kinetic handling methods and training. PhD Thesis, Department of Occupational Health and Safety, University of Aston, Birmingham.

Merriman, R.J., 1977. The role of audiometry in the prevention of occupational deafness. PhD Thesis, University of Aston, Birmingham.

Newbold, E.M., 1926. A contribution to the study of the Human Factor in the causation of Accidents. Report No 34, Industrial Fatigue Board, London.

Payne, D.E. and Barmack, J.E., 1963. An experimental field test of the Smith-Cummings-Sherman driver training system. Traffic Safety Research Review 7 (1): 10.

Perusse, M., 1978. Counting the near misses. Occupational Health: 123—126.

Perusse, M., 1980. Dimensions of perception and recognition of danger. PhD Thesis, University of Aston, Birmingham.

Pirani, M. and Reynolds, J., 1976. Gearing up for safety ... . Personnel Management: 25—29.

Powell, P.I., Hale, M., Martin, J. and Simon, M., 1970. 2000 Accidents. Report 21, National Institute of Industrial Psychology, London.

Raymond, S. and Tatum, S., 1977. An evaluation of the effectiveness of the RAC/ACU motorcycle training scheme. Road Safety Research Unit, Department of Civil Engineering, Salford University.

Rogers, R.W. and Mewborn, C.R., 1976. Fear appeals and attitude change: effects of a threat's noxiousness, probability of occurrence, and the efficacy of coping responses. J. Personality and Social Psychology 34 (1): 54—61.

Rotta, Sibour and De Gani, 1957. The frequency and gravity of accidents at work in a technical works, and in an apprentice centre. Rassemblements de Medecine Industrielle 26: 55.

Rubinsky, S. and Smith, N., 1973. Safety training by accident simulation. J. Applied Psychology 57 (1): 68—73.

Sell, R.G., 1970. Research report on safety propaganda. Paper submitted to Robens Committee on Safety and Health at Work.

Shaw, L. and Sichel, H.S., 1971. Accident Proneness. Pergamon Press, Oxford.

Shepherd, A., Marshall, E.C., Turner, A. and Duncan, K.D., 1977. Diagnosis of plant failure from a display panel: a comparison of three training methods. Ergonomics 20: 347—361.

Slovic, P., Fischhoff, B. and Lichtenstein, S., 1980. Risky assumptions. Psychology Today: 44—48.

Stammers, R.B., 1979. Simulation in training for nuclear power plant operators. Report 2. Ergonomrad AB.
Sulzer-Azaroff, B. and De Santamaria, M.C., 1980. Industrial safety hazard reduction through performance feedback. J. Applied Behaviour Analysis 13 (2): 287—295.
Surry, J., 1969. Industrial accidents: A human engineering appraisal. Department of Industrial Engineering, University of Toronto.
Tourigny, P., 1980. Etude des facteurs incitant au port des protecteurs auditifs individuels dans l'industrie du meuble. Master's Thesis, Department de Relations Industrielles, Université Laval, Quebec.
Tversky, A. and Kahneman, D., 1972. Availability as a determinant of frequency and probability judgements. Oregon Research Institute Technical Reports, 12 (1).
Van Zelst, R.H., 1954. Effect of age and experience upon accident rate. J. Applied Psychology 38: 313.
Weaver, F.J., Ramirez, A.G., Dorfman, S.B. and Raizner, A.E., 1979. Trainees' retention of cardiopulmonary resuscitation — how quickly they forget. J. American Medical Association 241 (9): 901—903.
Wilkins, P. and Sheppard, D., 1970. Attendance and audience reaction to a narrow road safety film. Technical Note TN 561, Road Research Laboratory, Crowthorne.
Winchell, S. and Safer, P., 1966. Teaching and testing lay and paramedical personnel in cardiopulmonary resuscitation. Anaesthesia and Analgesia 45: 441.
Zohar, D., Cohen, A. and Azar, N., 1980. Promoting increased use of ear protectors in noise through information feedback. Human Factors 22 (1): 69—79.
Zohar, D. and Fussfeld, N., 1981. A systems approach to organisational behaviour modification: theoretical considerations and empirical evidence. International review of Applied Psychology 30 (4): 491—505.

A REVIEW OF THE TRAFFIC SAFETY SITUATION IN SWEDEN WITH REGARD TO DIFFERENT STRATEGIES AND METHODS OF EVALUATING TRAFFIC SAFETY MEASURES

GÖRAN NILSSON

Swedish Road and Traffic Research Institute (VTI), S-581 01 Linköping (Sweden)

ABSTRACT

Nilsson, G., 1984. A review of the traffic safety situation in Sweden with regard to different strategies and methods of evaluating traffic safety measures. *Journal of Occupational Accidents*, 6: 35—47.

In industrialized countries road traffic accidents cause the majority of accidental deaths. The experience from accident and casualty investigations and studies of the effect of traffic safety countermeasures can to some extent be of great value in research on occupational accidents.

This paper illustrates problems connected with the evaluation of the effects of countermeasures, problems with accident and casualty statistics, examples from investigations and of the evaluation process in order to describe the benefit of investments to increase road traffic safety in relation to traffic safety strategies in Sweden.

BACKGROUND

Experience from traffic safety research may, in many cases, be of great value for the development of occupational accident research. In literature which describes risk concepts and risk analysis, examples are often taken from road traffic in order to explain in simple terms the concept of risk. These examples more often form an expression of the author's own view of traffic than a scientifically based evaluation.

In reality, road traffic is at least as complicated as any other situation where man constitutes a meaningful part of a system. If road traffic were under the administration of a single employer and road users were regarded as employees, this area of activity, if regarded as an occupational activity, would be the first to take action due to accidents and casualties. However, since the greater part of the road transport sector has developed into an individual activity and is regarded as such, it is the road user who has to bear the larger part of the responsibility for the negative consequences together with the police and hospital services.

As in many other situations the negative results of road traffic in the form of accidents and casualties can be regarded as a lottery. We know with considerable certainty how many road users are killed or injured annually but we cannot predict the time, location or consequences of each accident.

We are forced to describe traffic safety for groups of road users, parts of the population or sections of the road and street network before we adopt a plan of action which in general is suited to the corresponding structure.

Through various channels for recording accidents and casualties — the police, hospitals and insurance companies — it is possible to follow trends in the number and type of accidents from one period to another in various environments.

However, these sources produce distorted pictures of the traffic safety situation — no authority has the responsibility for reporting the correct level or providing a representative picture of the traffic safety problem. The majority of decisions are based on the number and severity of accidents which have already occurred. Risk assessments are beginning to draw attention but are most often based on historical knowledge.

ACCIDENT TRENDS, COUNTERMEASURES AND RISK CONDITIONS IN ROAD TRAFFIC

The rapidly increasing number of accidents during the 1950s was seen as a considerable social problem and was met with increased action by society in the form of investment in roads and streets, primarily to cope with the expansion of road traffic.

During the 1960s traffic regulation was given priority in the hope of limiting the continued increase in the number of accidents and casualties. The result was that traffic regulations — in the first instance, speed restrictions — reduced the consequences of accidents to a greater extent than the actual number of accidents. During the 1970s this led to the political possibility of action to reduce the consequences of accidents, which resulted in legislation on the use of seat belts and crash helmets.

After the Second World War, the number of accidents and casualties corresponded fully to the increase in car traffic until the middle of the 1950s. During the 1960s, the increase in accidents was on average about 5% units lower annually than the increase in car traffic due to the positive effect of different countermeasures.

The 1960s were characterised by over-use of cars and excessive travel at high speeds. The change to right-hand traffic in September 1967 probably meant a longterm positive effect on traffic safety — drivers and other road users started a process of re-education and speed limits were introduced throughout the whole road network, which contributed to reducing travel at high speeds.

Corresponding patterns are found in most industrialized countries but, for some reason, the traffic safety situation in Sweden has never become a "catastrophe", as it has, for example, in Japan, France or Finland. The explanation may lie in the conditions accompanying the change to right-hand

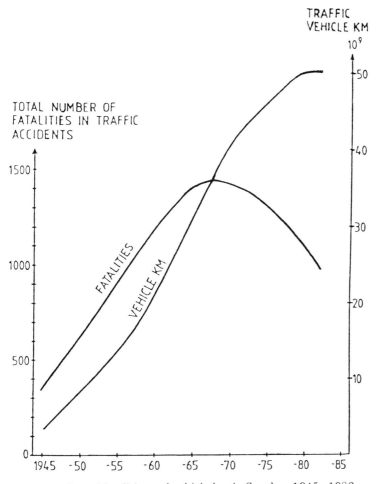

Fig. 1. Number of fatalities and vehicle-km in Sweden, 1945—1982.

traffic and roadbuilding policy during the 1960s and the years after the change. It is worth noting here that the planning of housing estates since 1965 has also taken full account of traffic safety requirements.

On the basis of information on the number of casualties in accidents reported to the police and information from travel habit surveys, it has been possible to quantify the risks to various road user categories on a national level (Thulin, 1981).

In Fig. 2, the number of casualties (including those killed) is related to the number of person-kilometres for various road user categories. The figure shows that high risks are associated with small degrees of exposure and low risks with high degrees of exposure, expressed in person-kilometres.

In Fig. 3, the number of casualties is related to the number of hours in traffic for various road-user categories. The difference in travel speeds between different road-user categories results in increased risks for motor

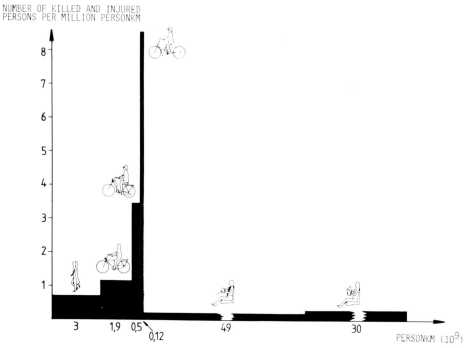

Fig. 2. Number of killed and injured people per million person-km for different road user groups.

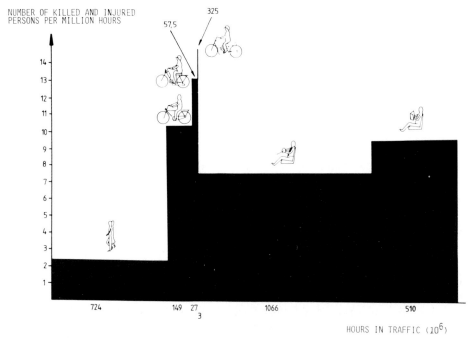

Fig. 3. Number of killed and injured persons per million hours in traffic for different road-user groups.

vehicles, especially for car drivers. Note that the areas shown are the product of risk and exposure, which is the same as the number of casualties.

In Fig. 4, time in traffic has been replaced by the number of trips without changing the internal order of the risk levels. The explanation for this is that the average travelling time per trip regardless of mode, is relatively constant (20 min.).

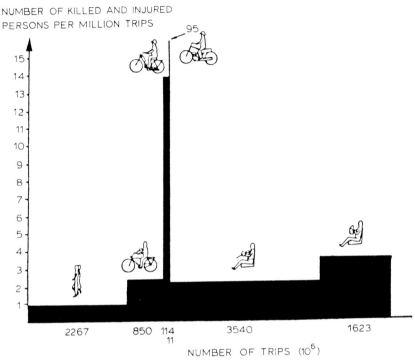

Fig. 4. Number of killed and injured persons per million trips for different road user groups.

ACCIDENT AND CASUALTY STATISTICS

Traffic safety descriptions based on reported accidents represent about 25% of all traffic casualties. The statistics are, however, almost comprehensive in the case of those killed or very seriously injured.

If we also include single accidents with bicycles and mopeds we find another 25% of the casualties. The majority of these injuries are recorded only by hospitals.

The remaining 50% of all traffic casualties consists of pedestrians who for some reason are injured when walking along roads or streets, hurrying to a bus, stumbling or falling or being injured when travelling by bus or tram.

The official traffic casualty statistics are still oriented towards describing accidents involving motor vehicles. This means that we receive a distorted picture of, for example, the need for hospital resources on the basis of police information, partly because of the fact that the majority of those killed do not need hospital treatment, while a large part of society's resources are required in order to care for cyclists, moped riders and pedestrians injured in single accidents not reported to the police.

Figure 5 shows a comparison between the official statistics for the severely injured and the corresponding information from hospitals (Nilsson and Thulin, 1982). Hospitals record a traffic safety problem which is twice as large as that recorded by the police and, in certain cases, is even larger in those age groups which use bicycles or walk more than other age groups.

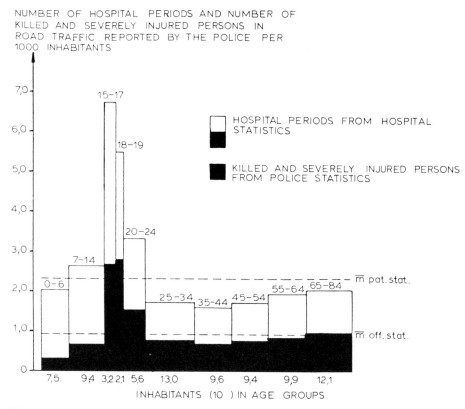

Fig. 5. Number of hospital periods per 1,000 inhabitants for people injured in road traffic, 1977. Number of people killed or severely injured in police-reported accidents per 1,000 inhabitants, 1977.

THE TRAFFIC SAFETY SITUATION AND THE EFFECTS OF DIFFERENT COUNTERMEASURES

Each countermeasure or change in traffic may alter the traffic safety situation in different ways. In principle, three different effects can be distinguished and recorded (Nilsson, 1981):

- Change in the exposure to accidents or personal injuries.

- Change in the risk of accidents or personal injuries.
- Change in the consequences of accidents or personal injuries.

This can be illustrated in a three-dimensional figure as in Fig. 6.

If we introduce a countermeasure this may influence one, two or all of the above effects. One example is that of speed limits which restrict exposure, risks and consequences. Increased fuel prices may influence exposure, better road standards may reduce risks and wider use of seat belts may reduce the consequences of accidents. This can be illustrated by altering the volume in Fig. 7 which is a direct measure of the number of traffic casualties.

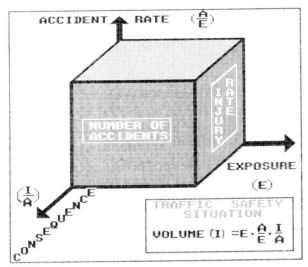

Fig. 6. Description of a traffic safety situation in terms of accident rate, accident consequences, exposure, number of accidents and injury rate.

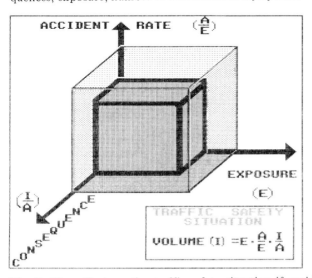

Fig. 7. The effect on the traffic safety situation if accident rate, accident consequence and exposure are reduced.

PROBLEMS IN EVALUATING THE EFFECTS OF COUNTERMEASURES

Table 1 surveys countermeasures and expected effects related to road environment, vehicles and road users, through a two-dimensional classification of road safety measures.

One example of a countermeasure for reducing exposure to accidents or the number of casualties is the separation of pedestrians from motor traffic by pedestrian tunnels or signals. A problem occurring here is that risks increase noticeably for the group of pedestrians who do not use the tunnels or who cross against a red light. If this group increases and becomes a majority, the difference in risk between crossing against a red light and crossing with a green light will decrease. However, this does not mean to say that it is safer to cross against a red light, but the expected effect of the countermeasure will be greatly reduced.

TABLE 1

Two-dimensional classification of road safety measures

Safety characteristic changed \ Contributing factors affected by the measure	Road user	Vehicle	Road/street
Exposure	Separation of different road user categories. Change in the number of trips.	Change of vehicle mileage. Regulation for vehicle traffic.	Regional planning of traffic control. Road improvements reducing travel distances.
Risk	Improvement of education, information and road user behaviour in relation to traffic rules.	Crash avoidance measures: Vehicle speed regulations. Vehicle equipment (studded tires, running lights in daytime, etc.).	Improvements on roads, streets or traffic network. Speed regulation. Road/street maintenance.
Accident consequence	Individual protection equipment (seat belts, helmets). First aid education.	Improved crash tolerance. Vehicle speed regulation.	"Soft" road side design. Alarm telephones.

The same problem appears when road standards are improved. There is, for example, no difference in accident risk for single accidents on motorways and on normal two-lane roads due to the higher speeds on motorways, or other factors. Since traffic in different directions on motorways is separated the risk of collisions is reduced. However, in those cases where a vehicle crosses the central reservation or travels in the wrong direction by mistake the consequences are often catastrophic because of the high speeds.

Apart from speed restrictions, countermeasures to reduce risks include other types of traffic regulation, road environment, vehicle-oriented measures and information and education. Vehicle-oriented measures may relate to brakes, tyres and lighting. A concrete example is that of vehicle lighting under daylight conditions ("running lights") and reflectors on bicycles and pedestrians.

Measures such as road markings and reflective posts along the road do not, in many cases, provide the expected effect since speeds increase. This means that an expected decrease in the accident risk is reduced or does not occur at all, at the same time as the consequences of accidents increase.

Action to reduce the consequences of accidents is not yet complete. Energy-absorbing modifications to various parts of vehicles have shown positive developments. Seat belts are not yet used by all occupants, especially those who perhaps would gain the most from using them, for example young drivers and passengers during nighttime. Pedestrians who are run over often receive unnecessary injuries as a result of the design of vehicles. It is precisely in quantifying how consequences are changed by various measures that most research is needed. This must be coordinated with the information available from hospitals.

TRAFFIC SAFETY RESEARCH

When investigating the effects of countermeasures there are probably few areas which have attracted as much interest as road traffic, at the same time as research resources have been limited.

In spite of the relatively comprehensive investigations and the fact that research has largely been done on the basis of available information not always suited to the problems studied, a considerable amount of knowledge has been gained regarding the background to traffic accidents. This is however often confused with ideas which have no connection with reality and with results which reflect rather an unsatisfactory investigation methodology rather than the effect on traffic safety. These "false truths" play a part which is not insignificant when decisions are made in the road transport sector.

There is, therefore, a body of opinion which has not wholly accepted the speed restriction system and is above all an opponent to further reductions in speed limits on roads where most accidents occur. Arguments that one becomes tired when driving at 90 km/h or that it is difficult to overtake where speed limits are in force are still heard occasionally.

Perhaps the greatest problem is that the effect of many road and traffic measures and of information and education activities is overestimated by the use of unsuitable investigation methods or by the absence of possibilities of measuring any effects.

Investigations in the form of statistical experiments — objects selected at random for treatment — to quantify the effects of measures are not accepted although practical experiments involving such measures are made without either checks being made or the desire to obtain information on the real effect of the measures.

The whole traffic system can be regarded as a gigantic experiment — new vehicles and new traffic regulations are all too often introduced without either planning or the intention of evaluating the effects. The discovery is made too late that it would be desirable to quantify the effects, usually because an individual person's suspicion has been aroused.

This unplanned situation often results in retrospective before and after studies. However, in those cases where action has been taken because of high accident rates, which is the normal case, it is impossible to eliminate random effects from the effects of the measures — regression to the mean effect. This procedure also means that all control possibilities are eliminated and the control object used must consist of objects which differ from the test object in some essential respect.

Measures are often applied to junctions, drivers and vehicles as though they had an abnormally high number of accidents or a high accident risk. As a result of the problem of the regression to the mean, the number of accidents or the accident risk will be reduced even if no countermeasures are taken. Many lists of countermeasures often include more random effects than actual effects.

A before-and-after study of road intersections in Sweden which remained unmodified during the years 1972—1978 (Table 2), showed that a halving of the number of personal injury accidents may be expected to be due simply to regression to the mean (Brüde and Larsson, 1982). It is not unusual for effects of this order of size to be reported in before and after studies of countermeasures.

Finally, I shall give an illustrative example regarding road conditions and traffic safety in winter (Andersson, 1978). Ice and snow on the road are a

TABLE 2

Regression to the mean injury accidents at intersections — Sweden (Figures in ( ) are exposure adjusted)

Number of intersections in group	No. of accidents per intersection during "before" period	Average no. of accidents per intersection during equivalent "after" period	Change (%)
2039	0 (0)	0.19	Increase
441	1 (0.85)	0.42	(−51)
119	2 (1.70)	0.71	(−59)
24	3 (2.56)	1.33	(−48)
14	4.143 (3.53)[a]	1.50	(−57)

[a]These figures are for sites with 4 or more accidents during 1972—1975.

problem for all road users. The occurrence of ice and snow can be reduced by spreading salt, which is expected to bring down the number of accidents. Figure 8A describes a situation where half the traffic uses roads affected by ice and snow while the other half uses roads free of ice and snow during a given period of time.

The accident rate is twice as high under conditions of ice and snow compared with roads free of ice and snow. Two thirds of the accidents occur under conditions of ice and snow.

If salt is used it is possible to eliminate half the ice and snow conditions.

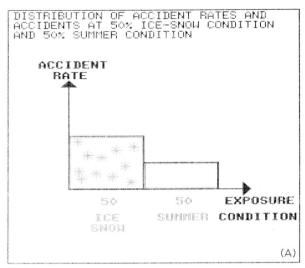

Fig. 8a.

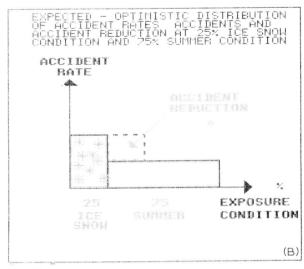

Fig. 8b.

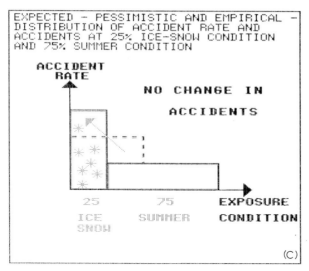

Fig. 8c.

Assuming that the accident risks are unchanged the result shown in Fig. 8B is obtained.

The investigations show, however, that reality is different (Fig. 8C).

The risk has increased by 50% under ice and snow conditions, which means that the number of accidents is unchanged. These results have led to an experiment being made with unsalted roads in Sweden, although on a modest scale. The results indicate that when the existence of ice and snow is reduced the accident rate increases on the remaining parts of the roads with ice and snow.

COST—BENEFIT ANALYSIS IN THE EVALUATION OF THE EFFECTS OF COUNTERMEASURES

When undertaking cost—benefit analysis, the road transport sector forms an interesting area. Traffic planning often relates to the expected accident costs, travel time costs and vehicle costs when illuminating the benefit of a new road or road improvement. However, I will not enter here into the evaluation problems arising mainly from accident and travel time costs.

In order to demonstrate society's valuation, a figure of S. Kr. 3 million is used for every person killed, 400,000 for every person severely injured and 150,000 for every person slightly injured. In reality this means that almost half the calculated accident costs are associated with fatalities.

The cost—benefit analysis is used both to create a priority ruling between different road projects and in conjunction with different road maintenance strategies.

Measures which alter vehicle speeds are of special interest here, since all costs can be related to speeds. They also include the present speed restriction systems which indirectly form the background to the relationship between accident and travel time costs.

For society as a whole the above-mentioned cost—benefit models must, however, be supplemented with other unquantified effects, such as employment, regional politics etc., which means that priority rulings are often invalidated. I believe that the latter problems coincide to a great extent with problems of research into occupational accidents and the consequences of these accidents for society.

REFERENCES

Andersson, K., 1978. Chemical de-icing of roads — effects on road accidents. Report 145, VTI, Linköping.
Brüde, U. and Larsson, J., 1982. The regression-to-mean effect — Some empirical examples concerning accidents at road junctions. Report 70, VTI, Linköping.
Nilsson, G., 1981. Traffic safety in terms of accidents, injuries, risks and consequences — a multidimensional method for the description of traffic safety situations. Ninth IRF World Meeting, Road Design and Safety. Stockholm 1981.
Nilsson, G. and Thulin, H., 1982. A description of the traffic safety situation based on patient statistics from the National Welfare Board. Report 237, VTI, Linköping.
Thulin, H., 1981. Traffic risks for different age groups and modes of transport. Report 209, VTI, Linköping.

BEHAVIOURAL CONTROL THROUGH PIECE-RATE WAGES

CARIN SUNDSTRÖM-FRISK

Swedish National Board of Occupational Safety and Health, S-171 84 Solna (Sweden)

ABSTRACT

Sundström-Frisk, C., 1984. Behavioural control through piece-rate wages. *Journal of Occupational Accidents*, 6: 49—59.

Risk taking or unsafe acts are seldom followed by accidents. On the contrary, these behaviours are often associated with different kinds of rewards. When applying strategies for promoting safe behaviour, factors rewarding unsafe acts must therefore first be identified and taken into consideration. Piece rate is supposed to constitute one such factor.

This paper presents accident data before and after a transition from piece-rate to time-based wages in the Swedish forestry industry. A reduction in the accident frequency and severity rates is observed. Mechanisms mediating the reduction were: reductions in stress (fewer human errors) and benefits associated with risk taking and removal of an obstacle to safety efforts.

THE BENEFITS OF RISK-TAKING

It pays to take a risk. At least most of the time, since risk-taking* or unsafe behaviour is rarely followed by an accident. On the contrary it is often associated with different kinds of rewards. The unsafe way is often — or is experienced to be — the easier, less time-consuming, less strenuous, more comfortable and productivity-increasing way of doing things. This is a discouraging conclusion to be drawn when comparing unsafe and safe behaviour in different settings (Lindström, 1974; Sundström-Frisk, 1978).

Effects of motivational efforts to get people to behave in a safer way, e.g. safety information campaigns, competitions or tokens, could be reduced or shortlived if, simultaneously, other rewards are operating in favour of unsafe behaviour. By what means can you convince a person, who has been applying "unsafe" methods for 20 years without having an accident, to change to "safer" but less rewarding work habits? The only reward you can offer him is the probability that he will not be hurt in the future. This is no reward for him since he does not expect an accident anyhow from his unsafe behaviour.

---

*By risk-taking is understood an intentional choice of behaviour which is known — from accident statistics — to increase the probability of having an accident. The risk-taking behaviour is chosen although there is time and opportunity to choose a safer way of acting.

The best way to cope with this situation is to redesign the physical environment so that injuries will not follow from an unsafe act; i.e. to eliminate the risk. But this cannot be done completely. Due to economical and other reasons, many risks which could be eliminated through design will remain. So we have to rely also on behavioural modification to prevent accidents.

To get a person to do the right thing (from a safety point of view) is not solely a question of motivation. To do the right thing, the person must have knowledge about actual risk, how to detect it and how to cope with it. He must also have the skill and ability to perform the action demanded. That is, the situation must not demand actions that are beyond his capability.

The motivational approach is valid only for behaviour in situations where the person has the time and opportunity to make a decision.

PIECE-RATE WAGES AND ACCIDENTS

As already mentioned, when applying strategies for simulating safe behaviour, factors rewarding *unsafe* acts must be defined and taken into consideration. In Sweden, big efforts were made in the late sixties and early seventies to reduce accidents in forestry work and particularly in the cutting operation. Effects, as shown in accident statistics, were not as good as could be expected. There were several explanations given for this. One was that the prevailing piece-rate wage system rewarded risk-taking and functioned as an obstacle to safety efforts (Lidberg, 1972; Lindström, 1974).

In 1975 several thousand Swedish forestry workers went on strike, primarily to get rid of a 30-year-old, flat, straight piece-rate system. On October 1975 an agreement was reached to abolish the piece-rate work and introduce time-based wages. This change of wage system offered an opportunity to study whether the piece-rate wage system really was one of the reasons why the safety efforts in forestry had not produced the expected results.

The relation between wage system and accidents has been widely debated. Research is, however, scarce and the results contradictory. In a review on accident research up till 1967 by Surry (1971) three pieces of research are mentioned: Keenan et al. (1951), Pajowsky (1959) and Maggio (1964). Pajowsky and Maggio postulated an increased accident risk when working on piece-rate but found no correlation. Keenan postulated a negative correlation, i.e. a decreased risk, when working on piece-rate due to increased arousal and readiness of the individual. He also found no correlation.

Kronlund et al. (1973) reported an increased severity rate when working on piece-rate compared to time-based wage forms in the mining industry. Similar results have been reported by Aronsson (1976) in highly mechanized work and by Mason (1977) and Granstam (1978) in forestry work.

When contradictory results are obtained a reasonable explanation for the controversy is that the relation studied exists under certain conditions only.

Powell et al. (1971) found in their analyses of 2000 accidents some of these conditions. They observed that the relation between piece-rate work and accidents was dependent on the design of the piece-rate wages, the way productivity was measured and if the connection between the workers' input and the productivity output was perceptible to the workers.

A conclusion that can thus be drawn after reviewing the literature on this subject is that there is a valid relation between the wage system and accidents under certain conditions but not in general.

QUESTIONS TO BE ANSWERED

The purpose of the actual work was not only to look for a relationship but — if there was one — to try to identify the mechanisms mediating this relation. The questions to be answered thus were:

1. Was there a reduction in accidents after transition from piece-rate to time-based wages?
2. If yes, was the reduction a temporary and/or normal variation?
3. Is the reduction to be attributed to the change of the wage system or is it related to confounding factors such as safety activities, change of risk exposure (decrease of productivity)?
4. How was the reduction brought about? That is, what mechanisms mediate a relation between wage systems and accidents?

WAGE SYSTEMS BEFORE AND AFTER THE STRIKE

Before October 1975 cutters were paid according to a flat, straight piece-rate system. Rates were determined by local agreements in accordance with centrally compiled price-lists. These price-lists were based on time studies and negotiations between the parties. Demands for a monthly salary had been loudest in the northern part of Sweden. In the rest of the country opinion was less uniform. Following the strike there were, as a result, two different agreements for the cutters:

A. *Flat monthly salary* (introduced in the northern part of Sweden). Wage rates determined by type of work and by a system of merit rating. The factors taken into account in the merit rating system were: work experience, training, time employed, versatility, days present and time spent working.
B. *Basic salary plus productivity bonus* (introduced in the remainder of Sweden).
   (i) Time-based element (approx. 85%).
   (ii) Straight productivity related element (approx. 15%).
   Rates for the time-based element were determined by type of work and merit rating.

METHOD

The study included 422 cutters and 65 persons holding supervisory positions, from thirteen forest districts of five different companies. Both of the areas covered by the two different collective agreements were represented, as were different geographical regions.

Data on accidents, hours worked and productivity were recorded individually for the 422 cutters from one period before and one period after the change of wage system. The periods compared are from October 1973 to January 1975 and from October 1975 to January 1977 (Fig. 1).

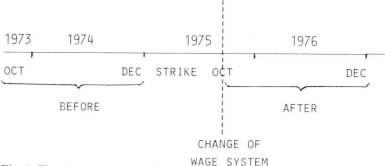

Fig. 1. The time periods studied.

A total of 360 of the 422 cutters and their supervisors were interviewed and/or answered a questionnaire covering questions about changes in:

- work behaviour in situations critical from a safety point of view
- experiences of physical and psychological stress
- productivity, both from a quantitative and qualitative point of view
- work organization
- safety efforts

Questions about work safety in the company and about changing of work organization and productivity were formulated in order to control for confounding factors. The assessment of productivity changes due to the change of wage system was — as expected — as difficult as assessing effects on accidents. This part of the study has been reported separately (Werner, 1979). The data collection was completed at the end of 1977.

RESULTS

The effects on accidents

A comparison before/after the change of wage system showed a reduction of 29% in accident frequency* (Table 1). The frequency during the first

---
*Number of accidents per one million working hours.

TABLE 1

Comparison between periods before and after the change of wage system for two different samples

This study (422 cutters)	Before	After	Difference (%)
Number of accidents per 1 million working hours	119	84	29
Number of sick-leave days per 1000 working hours	2.9	2.0	32
Swedish Forestry Service (2,500 cutters)			
Number of accidents per 1 million working hours	119	86	29
Number of sick-leave days per 1000 working hours	3.0	1.5	50

period was 119 and for the period after, 84. Furthermore the severity index** dropped by 32% from 2.94 to 1.99. The reduction in the frequency rate was less within the area of the flat monthly salary agreement compared to the area where there was a productivity bonus. The reduction in the severity index was the same in both areas.

Although our sample of cutters were more than 400, the *number* of accidents was relatively small — 137 for both periods together.

A comparison with accident statistics from the Swedish Forestry Service (which covers some 2,500 cutters) was made in order to test whether our results were representative. The reduction in the frequency rate was in this case 27% and there was a 50% drop in the severity index.

The reduction in accident frequency is similar in our sample to that of the Swedish Forestry Service. For the severity index there seems to be an understatement in our sample. Statistics for the cutting operation from the whole of Sweden reveal a considerable decrease — 23% — in the *number* of accidents occurring between 1975 and 1976 (from 3,898 to 2,916).

A reduction or a normal variation?

To answer this question we have to study the accident chart over a longer period. For this analysis we will once again use statistics from the Swedish Forestry Service. Statistics of the Swedish Forestry Service for the period 1971–1978 show that at no other time was there such a sharp fall in the accident frequency as that observed after the introduction of new wage froms in 1975 (Fig. 2). This drop thus could not be regarded as a normal variation. The same applies for the curve of the severity index (Fig. 3) where the drop between 1975 and 1976 is even more dramatic. The Swedish Forestry Service ceased to produce figures on the severity rate after 1976.

**Number of sick leave days due to accidents per 1,000 working hours.

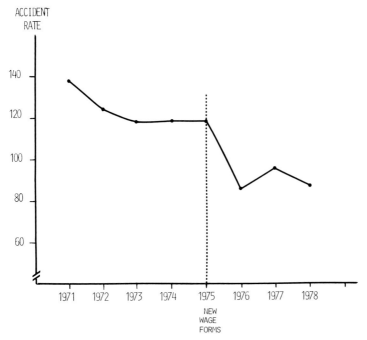

Fig. 2. Number of accidents per 1 million work hours during cutting operations for the Swedish Forestry Service, 1971—1978.

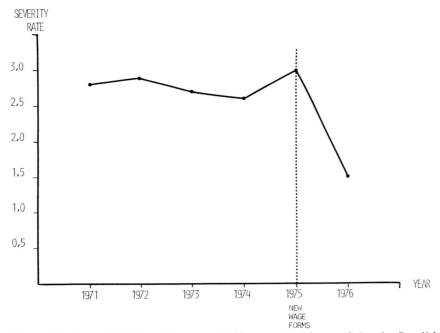

Fig. 3. Number of sick-leave days per 1000 hours of cutting work for the Swedish Forestry Service, 1971—1976.

To test whether the decrease shown in the Swedish Forestry Service statistics was temporary, we had to look for other sources of data. From the statistics of all forestry workers in Sweden we used another measure of severity index, i.e. number of sick-leave days *per accident*. This is a good measure to be used when describing changes in accidents since it is not biased by number of hours worked, nor by a change in productivity. Unfortunately there are no figures for cutters exclusively but for all types of forestry workers. As we can see from Fig. 4 there is a sharp fall after 1975 which continues to 1977.

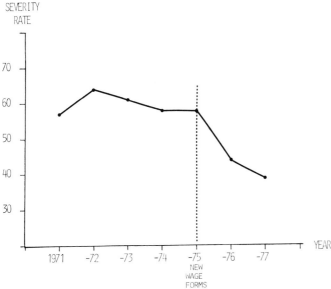

Fig. 4. Number of sick-leave days per accident in forestry work in the whole of Sweden, 1971—1977.

Confounding factors

During the seventies considerable effort was made to prevent accidents in forestry. Technical improvements of power saws (kick-back guards and chain brakes) and increased use of personal protective equipment — both activities reinforced by directions from the National Board — have had a positive effect on accidents. Training of the cutters, safety officers and safety stewards has also contributed. Complementary to all these activities, there has probably been an influence from public opinion and people's increased knowledge and consciousness of work environment issues.

The fall in the number of accidents occurring between 1971 and 1972 (Fig. 2) is largely attributable to the introduction of chain-braking kick-back guards on power saws and new safety directions for the felling operation.

In spite of the extensive safety activities, the accident rate remained fairly constant during the period 1972—1975. However, after 1975 the curve dropped dramatically. It is obvious that the abolition of straight piece-work has had an impact on this development.

At the same time as the wage system changed, wearing of leg protection was regulated by a direction from the National Board. Could this explain the reduction of accidents? If yes, there should be a considerable displacement of the distribution of injuries on different parts of the body towards a decrease of leg injuries and an increase of injuries on other parts of the body. Such a displacement is not to be found in the statistics.

TABLE 2

Distribution of accidents on different parts of the body

Year	Legs and feet (%)	Arms (%)	Others (%)
1975	29	11	60
1976	25	9	65

How to explain the reduction?

We have observed a reduction of accidents, but how is it brought about? From interviews and questionnaires we got three different explanations.

1. The benefits of risk-taking are reduced.

Cutters are no longer tempted to resort to faster, hazardous working methods in order to increase their earnings. By studying accident statistics we can pick work situations or operations which are very critical from a safety point of view. One example is bringing down lodged (hung) trees. In Sweden there is a special direction* that tells the cutter what working methods to be used in this situation. It also tells which methods should not be used, namely those which, according to statistics, are related to severe accidents. These methods are cross felling and taking down the bruised tree by felling the holding tree. We know from earlier studies that the cutters when working on piece-rate used the forbidden, unsafe methods because they saved time and energy (Lindström and Sundström-Frisk, 1976). It took about 14 minutes to use the right methods (if it was possible) and 1 to 4 minutes to use the forbidden methods. Obviously the cutter made money by using the risky ones.

In our study 77% of the cutters stated that they today have changed their behaviour when taking down hung trees. Either they are more careful and take their time for planning or they have changed to other methods.

The alternatives used today are e.g. asking a colleague for help (impossible earlier, since the colleague would lose income) or asking the tractor driver

---

*A regulation on working methods in the felling operation was issued in 1972 from the National Board of Occupational Safety and Health.

TABLE 3

Number of injuries when taking down hung trees (worker hit by the falling tree) from 1971 to 1980

Year	Number of injuries
1971	224
1972	157
1973	147
1974	105
1975	118
1976	38
1977	34
1978	--
1979	—
1980	36

for help. The change of work behaviour has, according to statistics, had a great impact. Accidents when taking down hung trees have dropped by 69% (Table 3).

2. Reduction of stress.

The second explanation for the decrease in accidents was reduction of stress. Man's performance is impaired when the organism is under excessive stress. Our ability and capacity to use our senses, i.e. register information and to make decisions and use our muscles become worse in a state of too high stress. The likelihood of error increases, as does the danger of accidents. All factors -- not just piece-work wages — which create excessive stress are bad from an accident standpoint.

In the interviews reduction of stress is the advantage of the new wage forms most often spontaneously mentioned, particularly stress that occurs in conjunction with production stoppages. Since the cutters experience less stress, they are also less likely to make errors of judgement or other mistakes, which also reduces the likelihood of accidents.

3. Piece-rate was an obstacle to safety efforts

Piece work has had a counter-productive effect on safety activities. Almost every one of the supervisors, cutter instructors, safety officials and other workers with safety duties mentioned during the interviews that it is now much easier to pursue safety activities than it was during the days of piece-work. Today the cutters give themselves time to listen to advice and instruction and to try new methods and equipment without having to worry about the fact that they might be losing money.

Follow-up of performance

In switching from piece-work to fixed wages, some decline in job productivity is to be expected. A simultaneous drop in accidents and productivity gives rise to the hypothesis that the number of accidents per unit logged is unchanged. This would imply no real change of the accident risk as a result of the new wage systems.

To be able to test this hypothesis, a follow-up study on productivity was made. We compared changes in accident frequency between cutters showing a great drop in productivity and cutters showing a slight drop or no drop at all. The analysis indicated no relation between drop in productivity and drop in accident frequency. The number of accidents in the analysis is, however, too small to draw the conclusion that such a relation does not exist. Obviously the cutters are exposed to higher risks if cutting 300 instead of ten trees a day. On the other hand, it is also obvious that the accident risk in cutting 300 trees is not 300 times as high as cutting one tree, since the risks are not equally distributed over all trees, all types of terrains, all types of climate, etc.

The answers from questionnaires and interviews indicate that only a small proportion of the drop in accidents can be explained by changes in performance. Nowadays, loggers allow themselves more time in situations that are critical from a safety standpoint. This affects productivity.

Furthermore, the decrease in the severity rate, defined as number of sick-leave days per accident, shows that a real improvement has taken place, which could not be explained by a decrease in productivity.

CONCLUSIONS

Statistical data on accidents in this study strongly support the idea that piece-rate wages have a negative impact on accidents. Furthermore, data from interviews and questionnaires provide logical and reasonable explanations of how this impact is brought about. This impact, however, is probably not valid for all situations; certain conditions have to be present. Forestry work is characterized by a high degree of manual operations, it is physically strenuous and there is a possibility to choose between different work techniques. The construction of the abolished piece-rate wages and the way of measuring individual productivity made the relation between actual work intensity and earnings obvious.

REFERENCES

Aronsson, G., 1976. From piece rate to monthly salary — Evaluation of a wage system change in highly mechanized work. Psychological laboratories, University of Stockholm (In Swedish).

Granstam, S., 1978. Accidents in forestry. Research report, Östra regionen, Domänverket (In Swedish).

Keenan, V., Kerr, W. and Sherman, W., 1951. Psychological climate and accidents in an automotive plant. Journal of Applied Psychology, 35: 2.

Kronlund, J., Carlssson, J., Jensen, I. and Sundström-Frisk, C., 1973. Democracy without power. The LKAB Mining Company after the strike. Prisma, Stockholm (In Swedish).

Lidberg, A., 1972. Accidents in the felling operation. Yrkesinspektionen, Umeå (In Swedish).
Lindström, K-G. and Sundström-Frisk, C., 1974. The assessment of safety directions. Undersökningsrapport AMP 101/74. Arbetarskyddsstyrelsen, Stockholm. (In Swedish).
Lindström, K-G. and Sundström-Frisk, C., 1976. Unsafe behaviour in the felling operation: Prevalence and controlling factors. Undersökningsrapport AMP 101/76. Arbetarskyddsstyrelsen, Stockholm (In Swedish).
Mason, K., 1977. The effect of piece work on accident rates in the logging industry. Journal of Occupational Accidents, 1 (3): 281—294.
Powell, P.I., Hale, M., Martin, J. and Simon, M., 1971. 2000 accidents. A shop floor study of their causes. Report nr 21. National Institute of Industrial Psychology. London.
Sundström-Frisk, C., 1978. Factors controlling unsafe work behaviour. Sveriges Skogsvårdsförbunds tidskrift, no 1-2, Stockholm (In Swedish).
Surry, J., 1971. Industrial accident research. A human engineering appraisal. University of Toronto, Toronto.

PSYCHOLOGICAL SAFETY DIAGNOSIS

U. BERNHARDT, C. GRAF HOYOS and G. HAUKE

Technical University Munich, Psychology Department, Lothstr. 17, Rg, D-8000 Munich 2 (FRG)

ABSTRACT

Bernhardt, U., Graf Hoyos, C. and Hauke, G., 1984. Psychological safety diagnosis. *Journal of Occupational Accidents*, 6: 61—70.

A questionnaire for preventative safety diagnosis of work systems, by means of which constellations of conditions conducive to hazards can be perceived, is being developed. The questionnaire consists of 250 items as well as interpretation aids, which can supply safety advisers in a factory with indicators of key problem areas in which unsafe working conditions are likely to develop. Suggestions for measures which can be taken to eliminate hazardous constellations of conditions in a factory are being developed.

Features of the work task, the work surroundings and the worker, all regarded as potential accident-causing conditions, are being studied. We attach special meaning to cognitive psychological requirements in hazardous situations as previous studies indicate that work sites with higher accident rates require a higher level of cognitive activity than do work sites with lower accident rates. Using individual case studies we intend to demonstrate how various components of the cognition process interact.

1. INTRODUCTION

Accident research to date is often criticized for having as its primary concern the interpretation of accidents. This retrospective analysis of accidents is, however, insufficient: the data found in accident reports do not suffice for the purposes of research. Recorded accidents, in any case, constitute only a tiny fraction of the total number of close calls and critical situations. Critics of accident research who argue from an ethical point of view complain that researchers do not begin their work until something *critical* has already happened; this complaint is also legitimate. We have therefore set ourselves the task of developing a preventive instrument, with the help of which indications of constellations that are critical to safety can be discovered. Our further goal is to develop measures by means of which such constellations can be avoided.

2. COPING WITH HAZARDOUS SITUATIONS

2.1. Point of departure for a diagnosis of safety in man — environment systems

An accident typically exhibits a multifactorial etiology, in which a chain of momentary inadequacies (e.g., misjudgement of distance, failure to hear a signal) and existing organizational weaknesses (e.g., insufficient consideration of suggestions for improvement) in a work system can negatively influence each other. No two accidents occur under identical conditions, even though the object involved in both accidents may be the same: a large number of possible interactions exists between the technical, organizational and personal realities in a situational context.

System reliability is only possible if a form of "adjustive behaviour" (McGlade, 1970), consisting primarily of anticipatory and compensatory behaviour, is required on the part of the subsystem "man". We have described this adaptability by means of a compilation of demands (e.g., the perception of unusual sound patterns) which a person must fulfill in order to demonstrate competent coping strategies and to reinforce the system goal of safety. These demands can also be seen as indications of the conditions which should ideally exist in a given system. Anticipatory behaviour here deals primarily with the reception and processing of information and with awareness, accompanied by the risk level accepted as a result thereof. The use of protective equipment, safety measures, etc. (Hoyos, 1980; Dawson et al., 1982) are indications of anticipatory behaviour. If a person accepts a certain risk level he thereby also defines the extent of his exposure to danger. Disturbances which occur in the system must be dealt with by means of compensatory behaviour. This behaviour takes the form of demands (e.g., deciding which preventive safety measures should be applied) and appropriate reactions (e.g., movements made to avoid danger).

Where then, among this set of anticipatory and compensatory demands, can a diagnostic point of departure be found? Results of previous research (Gockeln et al., 1981; Hoyos and Metzen, 1982) point to three criteria which can help answer this question.

The system goal of safety is endangered:

I. Certainly, by the existence of hazards of different kinds.

II. If the worker has (a) to cope with very exacting safety-critical demands or (b) to take into consideration simultaneously two or more safety-critical demands (e.g., keeping an eye on splinters produced by metal turning while trying to stay clear of a moving vehicle). We have coined the diagnostic term safety anomaly, a concept which shall include conditions (a) and (b) (Fig. 1).

III. It is often the case that demands of safe behaviour on the part of the subsystem "man" can be met quite realistically but the system goal of safety may be endangered because safe behaviour becomes obstructed by conditions of the context, e.g. noise of an approaching vehicle must be perceived (= a demand critical to safety) in a very loud environment (= context condi-

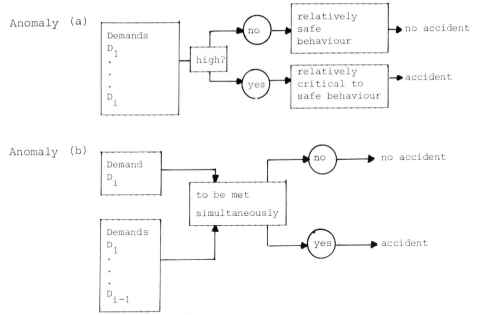

Fig. 1. Diagnoses of safety anomalies.

tion). In diagnostic terms, this situation (in which the demands on safe behaviour are incompatible with, e.g., the technical and organizational realities of the context), can be referred to as a constellation which is critical to safety (Fig. 2).

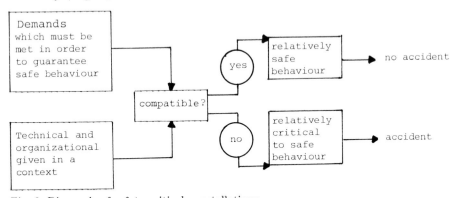

Fig. 2. Diagnosis of safety critical constellations.

2.2. Development of the "Safety Diagnosis Questionnaire" (SDQ)

The main goal of the required diagnostic process can be seen in the assessment of potential interactions between hazardous conditions and men which can lead to accidents, but about which little or nothing is presently known with respect to the likelihood of their leading to accidents. The best way to achieve this goal seemed to be a typical job analysis approach: the complete area of man — work interactions must be separated into job dimensions; job dimensions again must be split up in "work elements" each describing a certain demand on the worker, e.g., kind and amount of information to be processed, equipment needed for task accomplishment, environmental conditions. A representative example of this approach is the "Position Analysis Questionnaire" (PAQ) (McCormick et al., 1969; in German by Frieling and Hoyos, 1978) which served as a model for the SDQ. According to this model we had to make two steps: (1) establishing areas or dimensions of safety, and (2) formulating different demands on the worker in terms of safe behaviour as well as contextual conditions favourable or unfavourable for safe behaviour. This has been done on the basis of accident research data as well as of detailed case studies.

Empirical research on accidents has shown that risk-taking behaviour in work systems is dependent (with highly significant correlations) on human reception and processing of information and on the extent to which an individual is free to make his own choices (Hoyos et al., 1977). In a comparison between samples of industrial work sites with frequent accidents and samples with less frequent accidents, we have found that, in addition to the above factors, work structure, work equipment, the furnishings of the work place, communication, cooperation and the extent of control over behaviour affect the hazardousness of a situation (Gockeln et al., 1981). Häkkinen (1977), Carlsson (1980) and Tuominen and Saari (1981), using in part older paradigms, have produced similar or less detailed results in certain areas of the systems mentioned. Finally, we conducted detailed studies of individual cases by means of discussions with factory safety experts. This preliminary survey resulted in the formulation of the following relevant units or dimensions for the SDQ (Fig. 3).

```
1.  Structure and Implementation of Work Safety Programs
2.  Formal Organization of Work
3.  Environmental Influences
4.  Possibilities of Hazards and Safety in the Work System
5.  Presentation and Processing of Information
6.  Execution of the Work Task
7.  Communication and Cooperation
8.  Acting in Safety-Critical Situations
```

Fig. 3. Units of the SDQ.

By reasons of better administration of the questionnaire the criteria I–III are not represented by these units in the same order but it can be seen easily that unit 4 refers to hazards within the system in question, units 2 and 3 contain contextual conditions, and units 5–8 deal with different kinds of demands. Unit 1 gives an overall picture of the safety activities of an industrial setting. According to McCormick we have assumed that each man — environment system (e.g., work systems, traffic systems) can be broken down into observable and ascertainable elements and can thus also be fully understood. Adopting this approach we went on to formulate questions which we called Safety Elements (SE). Each unit of the SDQ (Fig. 3) comprises a series of Safety Elements.

To begin with, information is gathered about those demands which a person must fulfill in order to behave safely in a given system (Fig. 4).

Item 3.2.08 Recognizing smells

    How important is it for an employee to recognize smells as indicators of harmful energy (e.g., the smell of burnt rubber or oil, the escape of poisonous and dangerous gas)?

Item 5.2.11 Agreeing on use of space

    How important is it for an employee to come to an agreement with his colleagues concerning the utilization of space (e.g., storage spaces that are used by several employees)?

Fig. 4. Examples from the demand system.

Also, those technological and organizational conditions can be determined which may influence behaviour and can be considered as contextual factors affecting safe or unsafe behaviour (Fig. 5).

Item 1.2.03 Safety checks for the work site itself

    Are safety checks at the work site conducted in the factory (e.g., checks on quantities of dangerous (toxic) substances present, checking of technical safety equipment)?

Item 2.1.05 Danger caused by disorder

    How often do dangerous situations arise in the work area due to the unsafe placement of work materials (e.g., tools or other objects left lying around the work area or in the path of moving vehicles)?

Fig. 5. Examples of contextual conditions.

We came up with a collection of about 250 SE covering all units or dimensions listed in Fig. 3. To assure their intelligibility, safety relatedness, and above all, the practicability of the SDQ, 160 safety experts examined samples of these items and evaluated them on respective rating scales. In general, experts supported the approach governing the first version of the SDQ. Average ratings on a 5-point scale for these three criteria were evaluated for each item. However, more benefit could be gained from critical comments on certain items. They stimulated very much a thorough revision of the first version of the SDQ.

2.3. Case study: Safety diagnosis

We shall explain the nature of our questionnaire and of our diagnostic method by taking a look at a typical work system; we shall use case III to demonstrate how this diagnostic method can be applied to safety-critical constellations (Fig. 6).

Fig. 6. An industrial welder's place of work.

Even the inexperienced observer can clearly see inherent dangers here. High temperatures are required for welding (e.g., arc welding of metal containers) or using an oxy-acetylene cutter (e.g., cutting open of steel jackets). A welding flame can have a temperature of up to 3200°C; the temperature of an electric arc can exceed 4000°C. Welding sparks and drops even reach temperatures of 1200°C and higher. An experienced safety expert might shrug his shoulders at this point and say: "So what. Such is the life of a welder." Is it then sensible to make psychological diagnoses of this place of work?

A classification of various welding work sites using the SDQ resulted in the establishment of a set of factors for each section of the questionnaire. These yielded information which is useful in discovering the safety anomalies described in cases I and II (see Fig. 1); they will not be further discussed here. It is the discovery of incompatibility between certain demands and technical and organizational conditions that result in an especially interesting picture of such work sites and that allow a more detailed diagnosis of safety to be made (see case III in Fig. 2). The following are examples of two constellations which we have diagnosed as being critical to safety. The demands in these cases are in the areas of decision-making processes, communication processes and processes involving the execution of the work task. These are considered in the light of the technical and organizational conditions they are incompatible with (Fig. 7).

We will try to explain why these are cases of incompatible constellations, i.e., from a diagnostic point of view, constellations which are critical to safety.

CONSTELLATION 1: Decisions made during the work preparation phase, which are relevant to safety

DEMAND (Item-no.)	Incompatible conditions CONDITIONS (Item-no.)
8.2.01 Decision complexity	8.2.03 Deciding despite uncertainty
3.3.07 Consolidation of information	6.3.06 Time pressure
	2.3.06 Ignoring safety measures
3.3.05 Evaluation of mechanical stress conditions	8.2.04 Persons responsible for making decisions
3.3.06 Evaluation of spatial conditions	8.2.08 Delayed consequences of wrong decisions
	4.4.06 Use of borrowed tools
	4.4.05 Small working areas
	2.1.02 Secondary sources of energy
	6.1.08 Working in an unfamiliar environment

CONSTELLATION 2: Danger due to lack of cooperation

DEMAND (Item-no.)	Incompatible condition CONDITION (Item-no.)
5.1.10 Communication with changing cooperation partners	2.1.03 Risk area
	6.3.06 Time pressure
5.2.11 Agreement about utilization of space	5.2.17 Contradictory verbal instructions
	1.2.16 Contact with work safety personnel
5.1.01 Verbal communication	5.1.05 Working under unfavourable acoustic conditions
	5.1.06 Communication with foreigners

Fig. 7. Constellations critical to safety.

Constellation 1: The insidiousness of decision-making.
When a welder is in the process of preparing for his actual work task he is already making decisions (demand-item no. 8.2.01) which can have considerable consequences. He must actively seek information (demand-items 3.3.05, 3.3.06) and must integrate this information in accordance with fixed rules (demand-item 3.3.07). Above all, he must determine whether objects which can easily explode or ignite, including deposits of dust, are to be found in the vicinity of the danger zone. He is often under considerable pressure to work as quickly as possible (condition-item 6.3.02), e.g., when he is doing repair work, so that he has no choice but to drastically limit his exploration of the work area. This condition has an extremely negative effect on his ability to calculate the risk involved, especially since he must often work in surroundings he is not familiar with (condition-item 6.1.08); in such an environment a welder will usually not even know where to look for information that could be relevant to his work. If, e.g., he has to weld pipes which continue on through a wall into adjoining rooms, then he should know about the presence of explosive or ignitable material in those rooms, which might be inflamed by the secondary energy which is produced (condition-item 2.1.02). He sometimes does not seek this information because he is in an unfamiliar environment, cannot gain entry to the rooms in question and, in addition, does not have enough time. He thus makes decisions despite a considerable degree of uncertainty (condition-item 8.2.03). This is also the case when he sets up his gas cylinders and generally does not know whether or not there is a source of heat (e.g., a hot stove in the break room) nearby. If the work area is too small (condition-item 4.4.05), then the very act of setting up gas cylinders makes the work situation precariously unsafe (condition-item 8.2.03), because another employee, coming through a narrow aisle in a hurry, might easily knock over the cylinders and thereby cause great damage. One more important condition should be mentioned here, because it repeatedly and in a very specific way keeps workers from making decisions adequate to their safety at work: delayed consequences of wrong decisions are quite common at welding work sites (condition-item 8.2.08). Hot metal drops or sparks produced while the welding task is being performed can fall or fly through the work area and cause materials (layer of dust, e.g.) to smoulder in not easily detectable places. These are often overlooked and can sometimes cause fires which do not break out until after working hours. Since such a critical situation is not seen as being the direct result of a decision or risk calculation, a broader search for information which could lead to a correction of the inadequate perception of the situation and the resulting critical situation does not take place.

Constellation 2: Danger caused by lack of cooperation among employees.
An important demand states that a welder, in the course of preparing for his work task, should communicate with the personnel working in the

department in which he is to do repair work (item 5.1.10). This is often a question of coming to a verbal agreement (item 6.1.01) about the utilization of space (item 5.2.11). It is characteristic of such a demand that difficulties are inherent in the frequent need of the welder for a considerable amount of space (Fig. 8).

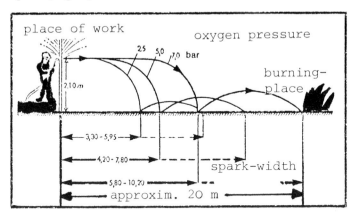

Fig. 8. Risk area.

As one can see, the risk area (condition-item 2.1.03) — here an area with a radius of up to 20 m in which fires can break out — can be a condition that makes agreement about the utilization of space very difficult to reach. The reason is that other employees who use this same area in the course of their work and who are also under pressure of time (condition-item 6.3.02) have to temporarily deposit their inflammable material (in a chemical factory, for example) elsewhere. The department supervisor believes that a welder should try to minimize the amount of work involved in each repair task because of the inconvenience involved for other workers; the welder himself, however, has a very different set of instructions, which he feels he has to follow. The contradictions inherent in these various instructions (condition-item 6.2.17) and the rigid schedule (condition-item 6.3.02) according to which he must work, keep him from trying to reach time-consuming agreements with other workers. The hazards which thus arise are apparent. Since the possibilities the welder has of coming into contact with work safety personnel are quite unsatisfactory, a fire patrol which might exist for the duration of the welding task and which thus might facilitate communication on the utilization of space cannot be organized. Two additional important factors which make the communication about the utilization of space in this special case virtually impossible are that the workers involved are generally foreign workers who understand little German and that communication must take place against a background of a high noise level (approx. 85 dB) (condition-item 5.1.05).

3. CONCLUSION

Constellations which are critical to safety are — as we have tried to show — accumulations of contradictory factors stemming from the most varied areas of the man—environment system. These as well as the safety anomalies do not necessarily lead to accidents but do make them more probable. It is very difficult to get a clear picture of the complex units of interaction which comprise these constellations. These can become recognizable and transparent, at least in part, by means of Safety Elements. The Safety Diagnosis Questionnaire and the diagnostic compatibility model could be helpful in trying to find non-trivial, not directly perceivable incompatibilities between the demands of a system and conditions for safe behaviour. This hope seems to be justified by the first 92 analyses we have done in different factories.

REFERENCES

Carlsson, J., 1980. Accident prevention in the steel industry. Arbetsolyeksfallsgruppen, Occupational Accident Research Unit.
Dawson, S., Poynter, P. and Stevens, D., 1982. Strategies for controlling hazards at work. Journal of Safety Research, 13: 95—112.
Frieling, E. and Hoyos, C. Graf, 1978. Fragebogen zur Arbeitsanalyse. Bern: Huber.
Gockeln, R., Hoyos, C. Graf and Palecek, H., 1981. Handlungsorientierte Gefährdungsanalysen an Unfallschwerpunkten der Stahlindustrie. Berichte aus dem Institut für Psychologie und Erziehungswissenschaft der Technischen Universität München, Lehrstuhl für Psychologie, Nr. 7.
Häkkinen, K., 1977. Crane accidents and their prevention. Journal of Occupational Accidents, 1: 353—361.
Hoyos, C. Graf, Keller, H. and Kannheiser, W., 1977. Risikobezogene Entscheidungen in Mensch-Maschine-Systemen: Arbeitsplatzanalysen in Industriebetrieben. Berichte aus dem Institut für Psychologie und Erziehungswissenschaft der Technischen Universität München, Lehrstuhl für Psychologie, Nr. 3.
Hoyos, C. Graf, 1980. Psychologische Unfall- und Sicherheitsforschung. Stuttgart: Kohlhammer.
Hoyos, C. Graf and Metzen, H., 1982. Belastung und Beanspruchung bei Steuerungs- und Überwachungstätigkeiten — Untersuchungen in Fahrdienstleitungen der Deutschen Bundesbahn. Berichte aus dem Institut für Psychologie und Erziehungswissenschaft der Technischen Universität München, Lehrstuhl für Psychologie, Nr. 9.
McCormick, E.J., Jeanneret, P.R. and Mecham, R.C., 1969. The development and background of the Position Analysis Questionnaire (PAQ). Rep. No. 5. Lafayette, Ind.: Occupational Research Center, Purdue University.
McGlade, F.S., 1970. Adjustive Behaviour and Safe Performance. Springfield, Ill.: Thomas.
Tuominen, R. and Saari, J., 1981. A model for analysis of accidents and its application. Tampere University of Technology, Dept. of Mech. Engineering, Lab. Protection, Report 15.

ABSTRACTS

Experience of Implementing Safety Information and Management Systems in Industrial Companies

M.T. HO

INRS, Dept. of General Studies, 30 rue Olivier Moyer, 75680 Paris, Cedex 14 (France)

To introduce and operate a safety information system or, in more general terms, a safety management system in a company is one of the central problems in occupational accident prevention.

For some ten years now INRS has developed and put into practice in industry a method for analysing occupational accidents, choosing preventive measures and following them up.

The report describes the INRS system and how it has been introduced into French firms. Two different systems which have been successfully implemented in two firms are presented for comparison.

The most significant conclusions are: Different systems (or strategies) can give good results if they are implemented (or applied) with conviction. As well as this question of motivation it is important that they should be sufficiently broad and complete in their scope. A simple accident information system is generally inadequate; it has to be integrated into a wider system including methods and structures for defining, adopting and following up preventive measures. In other words, once it has been introduced, an information system needs, for its survival, to be backed up by practical results.

A Behavioural Approach to Work Motivation

JUDITH L. KOMAKI

Purdue University, Psychological Sciences, West Lafayette, Indiana 47907 (U.S.A.)

The motivation for workers to perform safely is missing in most work settings. The applied behavior analysis approach is a particularly suitable strategy because of its focus on performance consequences as a source of motivation. Three recent field experiments illustrate how desired performance is taught, appraised, and reinforced. The last two studies also demonstrate the critical role of performance consequences in work motivation.

Alcohol and Fatal Work Accidents

ELISABETH LAGERLÖF

National Board of Occupational Safety and Health, S-171 84 Solna (Sweden)

MILAN VALVERIUS

State Institute of Forensic Medicine, Stockholm (Sweden)

PETER WESTERHOLM

Swedish Trade Union, Stockholm (Sweden)

292 fatal work accidents from 1979—1982 were compared with the registers at different Swedish Institutes of Forensic Medicine. From the autopsy reports at the medico-legal districts, data from examinations for drugs and alcohol were collected. Also data on the state of the liver (fatty liver or cirrhosis) were examined.

In 50 cases no specimens for alcohol analyses were taken. Testing for drugs was only done in 45 cases.

In 7 cases, or 2.9% of the investigated fatal work accidents, positive blood alcohol levels were found. This can be compared with accidents in traffic, where 46% of the drivers in single car accidents and 7% in head-on collisions had alcohol blood levels over the legal limit. No positive test for drugs was found. 19.2% had changes in the liver as compared to 28.9 in another studied material.

The conclusions are that fatal accidents at work only rarely are associated with positive blood levels of alcohol. Also there is no indication that alcohol abusers are more often involved in fatal work accidents than non-abusers. The data does not support the hypothesis that drug use plays an important role in work accidents.

A Problem-oriented, Interdisciplinary Approach to Safety Problems

JOHN STOOP

Delft University of Technology, Postbus 5050, 2600 GB Delft (The Netherlands)

An example is given of how we use the schematic approach as developed in this paper.

The problem concerns the occurrence of hand and finger injuries due to the use of pruning shears in the fruit growers' trade. A step-by-step approach is elaborated for the several stages of the problem. An explanation is given of the use of the intersections, the matrix, the functional demands, the selections of most desirable solutions, the demands for technical redesign of the pruning shears and the design of protective armoured gloves.

The method gives a fair insight into the possibilities, limitations, and consequences of the solutions chosen.

Interviewing Employees about Near Accidents as a Means of Initiating Accident Prevention Activities: A Review of Swedish Research

NED CARTER

National Board of Occupational Safety and Health, S-171 84 Solna (Sweden)

Swedish research on interviewing employees about their involvement in near and minor accidents as a means of initiating accident prevention efforts is reviewed. The object of the review is to summarize and evaluate this research to determine if the methodology can provide an important contribution to safety activities. An attempt is made to identify factors which increase the likelihood of a successful application of the methodology by local safety officials. Periods of interviewing are generally associated with improvements in safety consciousness and an increase in safety activities. Findings indicate that interviewing employees about their experiences can be a useful complement to local safety activities.

Developing Routines in Efforts to Prevent Occupational Accidents: An Accident Investigation Group

EWA MENCKEL

National Board of Occupational Safety and Health, S-171 84 Solna (Sweden)

Information about accidents which have occurred should be put to maximal use in local safety activities. In Sweden, supervisors are frequently delegated responsibility for accident reporting but are rarely trained in accident analysis skills. In the present study, a special group was created to assist supervisors in their accident investigations. During the first year, accidents were reported more promptly, more information was obtained, accident prevention activities increased and there was a reduction in accident frequency and severity as compared to previous years. The number of near accidents reported also increased dramatically. Follow-ups after two and three years confirmed the previous results.

VALIDITY AND UTILITY OF THEORIES AND MODELS OF ACCIDENTS

OCCUPATIONAL ACCIDENT RESEARCH AND SYSTEMS APPROACH

JACQUES LEPLAT

Laboratoire de Psychologie du Travail de l'E.P.H.E., 41, rue Gay-Lussac, 75005 Paris (France)

ABSTRACT

Leplat, J., 1984. Occupational accident research and systems approach. *Journal of Occupational Accidents*, 6: 77—89.

Effects of the use of the systems approach in accident research are discussed. Any accident may be analyzed in terms of a variety of systems and it is therefore always necessary to specify the nature and the interrelations of the systems to be studied.

Four types of research based on the systems approach are presented. They fall within the framework of:
(1) Accidents as the expression of systems dysfunctioning; identification of the classes of dysfunctionings.
(2) Relationship between the accident and other indicators of dysfunctionings.
(3) Accidents and system changes.
(4) Systems of reference of accidents in individual appreciation.

The possible contribution of systems analysis to furthering advances in the field of accident research are assessed.

INTRODUCTION

The theory of systems has inspired many scientific disciplines which, however, quite often just borrowed some general concepts from it. For example, only a few scientists working in the field of the psychology or sociology of organizations used the formalized aspects of this theory even though they were influenced by what some researchers (Emery, 1969) have called "systems thinking". Some scientists approached the study of sociotechnical systems with this new mode of thinking (Emery et al., 1960). The present analysis of work accidents is also done within this perspective. This short paper will not specifically mention the other system approaches which, in fact, consider only part of the socio-technical system — for example, industrial engineering and human factors. This more complete system integrates the partial technical and human sub-systems and their interactions.

On such a large subject, it is only possible to emphasize some points. I shall first mention some characteristics of systems which, in my view, should orient studies on accidents. I shall then present some studies conducted within the system perspective. Finally, I shall show how the study of accidents can benefit from such a perspective.

THE SYSTEM PERSPECTIVE IN STUDIES ON ACCIDENTS

There are many definitions of a system, each one emphasizing the features considered as typical. Ashby's (1970) definition is one of the most general: he considers that to define a system is "to list the variables that are to be taken into account" (p. 40). This apparently banal definition underlines that it is indispensable in every case to list precisely the variables or elements constituting the system considered. A system is indeed always a model, an abstraction conceived by the analyst. Within the same object one might distinguish different systems of variables, depending on the aim of the study; analogously, the boundary of a system, what separates the system from its environment, has to be defined. A system is never given: it has to be built and this construction is often the most difficult part of a study.

Sociotechnical systems, man-machine systems, also named "living organization" (Faverge et al., 1970) which I shall often mention here, are open systems; their elements or variables are both human and technical. These systems are very diverse, from the elementary system constituted by the individual work place and possibly its components, to the whole plant, the workshop, the production unit, etc. One may also become interested in the social systems constituted by variably sized sets of individuals: the team, the department, a professional group.

It is important to note that in any sociotechnical system, whether elementary or complete, human elements have a particular status. This particular status is expressed by the fact that individuals may consider their activity as serving aims that do not necessarily coincide with the aims of the whole system. Individuals and groups may redefine their goals as well as the means of attaining them (Hackman, 1969); hence, the distinction between prescribed task and real task (Herbst, 1974; Leplat et al., 1983) is quite necessary.

Within this perspective, what is an accident? Before answering this question, I want to make it clear that an accident will be considered as an immediate injury to the bodily integrity of the human element in the system. The notion of immediate injury permits differentiation between accidents and long term disorders (deafness, professional diseases . . .) caused by nuisances due to working conditions. An accident is an undesired – hence unplanned – consequence of the system functioning (what is possibly planned is prevention and/or remedial measures). An accident is therefore a consequence of a dysfunctioning in the system which does not work as planned (Leplat et al., 1979). Consequently, the study of accidents may be envisaged as the study of the dysfunction(s) that caused them.

Conceiving an accident in terms of a symptom of a system dysfunctioning has several consequences:

(1) It provides an objective in the study of accidents, i.e. the identification of the dysfunctionings, and an objective for safety measures, i.e. the reduction of the dysfunctionings. It is no longer a question of acting only upon the accident symptom (symptomatic treatment) but rather of acting upon the dysfunctions that caused it (curative treatment).
(2) It leads to the investigation of various symptoms of the same dysfunctionings, i.e. other accidents or incidents which constitute additional indicators of danger.
(3) Accidents may be studied by reference to several types of system: the composition of the considered systems, the relations between them, and their relative importance should be defined very precisely.
(4) It runs counter to a simplified causal conception of the origin of accidents.
(5) When accidents are considered as the result of the system's functioning, one is led to investigate the mechanisms of their production (one might say of their genesis).

TYPES OF DYSFUNCTIONING

Within the system perspective, an essential task is to identify and classify dysfunctionings. The notion of dysfunctioning is to be compared with the notion of risk factor or potential factor of accident. It should also be noted that one may trace back from the dysfunctioning to its origin. The same event may be considered as a dysfunction when paying attention to its origins, and as a source of dysfunctioning when paying attention to its consequences. With Cuny (Leplat et al., 1979), I analyzed in detail different categories of dysfunctionings. These categories were mostly defined on the basis of studies conducted in the iron and steel industry (X., 1969). Some of these categories will now be presented.

1. Deficiencies in the articulation of subsystems

A system may be divided into subsystems. The functioning of the global system depends not only on the functioning of these subsystems but also on the manner in which these functionings are coordinated to meet the aims of the global system. Several dysfunctions generating accidents which are ascribable to failures in this coordination can be mentioned.

(a) Boundary areas as zones of insecurity
When considering departments of a factory as types of subsystems, boundary areas between departments often constitute zones of uncertainty in which the functions of each department are poorly defined; thus, they constitute zones of insecurity.
As an illustration, there is the fact that, in an iron and steel plant, frequent accidents and incidents occurred in places of evacuation of blast

furnace products. It was found that these places were at the boundary of the blast furnace department and the transport department. The distribution of the functions to be exercised in these places not being satisfactory, this gave rise to conflicts. For example, (1) the floor was not cleaned properly because each department considered the other to be responsible for cleaning, (2) a signal informing transport workers of the state of the blast furnace did not work and was not repaired because the members of each department were waiting for the members of the other department to repair it.

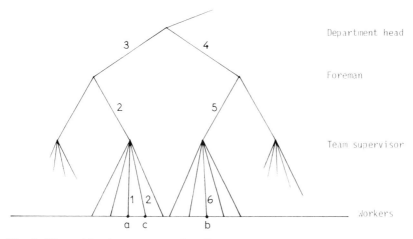

Fig. 1. Hierarchic chart representing distances between individuals (the distance between individuals a and b is 6 since there are 6 transitions from one hierarchical level to another when information passes from a to b; between a and c there are only 2).

Faverge (1967) proposed interpreting some dysfunctionings observed in these boundary zones in terms of the hierarchic distance of the workers to the common executive. This distance is evaluated (Fig. 1) by the number of lines of the hierarchic chart separating one individual from another (or by the number of transitions from one hierarchical level to another). The greater this distance, the more difficult the communication and the higher the level of uncertainty and insecurity.

There are also boundary zones between other subsystems, for instance, between teams, operators, professional groups, and it would be easy to find examples of studies revealing that accidents occurred in these zones.

(b) Zones of overlapping as zones of insecurity

Systems present overlappings. One of the many definitions of overlapping underlines that the same function is realized by the cooperation of two subsystems, another that the activity of two (or more) systems is exerted on the same site. I shall deal only with the second case and refer to it as co-activity. At the deaprtment level, this case is illustrated by the presence on the same site of a construction or maintenance department which is external to the plant and of a department belonging to the plant. The activities of the two departments often seem to be poorly coordinated; for instance, the safety rules observed by one department are ignored by the other. An investigation (X, 1969) in the steel industry (transport department and external construction service) revealed that 67% of technical incidents with material damage occurred in zones of co-activity representing only a small amount of the zones assigned to transport. Such conditions are frequent in modern industry which often uses external enterprises to deal with some aspects of maintenance, repair, cleaning, building, etc. The organization of situations of co-activity is an important task for safety.

It should be noted that situations of co-activity may also be found at the interindividual level.

(c) Asynchronous evolution of the subsystems of a system

Changes are sometimes introduced into subsystems, due to the modernization of some equipment. Such changes are often carefully designed but their consequences on other parts of the system are sometimes totally neglected. For instance, a production procedure is modified in order to increase production greatly, but no change in the means for removing the products is planned, leading to pile-ups, and to temporal constraints on workers, both factors which impair safety.

Cases in which a properly designed part of a system deteriorates may also be considered as belonging to this class of dysfunction. For instance, a poor stretch in an otherwise good road or a faulty element in a new system. In these cases, the erroneous expectations of the user, the inaccurate appreciation of the situation, may lead to accidents.

(2) Lack of link-up between the elements of a system

The quality of the link-up between the elements of a system is not only the condition of efficiency but can also be the condition of safety. This phenomenon may be observed at different levels of the systems, as illustrated by the following examples.

The link-up may concern firstly the members of a group: in this case, team or group cohesion. Several studies gave evidence of the relationship between lack of cohesion and safety (X., 1969). This relationship may be explained because a lack of cohesion is often an obstacle to the circulation of information within the group; information is indispensable when the actions of an individual depend on what another individual says or does.

The link-up may also concern the relations between the individuals and the material elements. The quality of this link-up is determined by reference to the task. When the capacities of the individual do not correspond to the task requirements, the result is quite often impaired safety (Johnson, 1980). However, the reverse is equally true. For instance, links between the displacements of a signal and the moves of a command which do not follow the sensorimotor stereotypes of users may lead to accidents. The notion of compatibility in ergonomics provides us with very useful data on this issue.

ACCIDENTS AND OTHER CLUES TO DYSFUNCTIONING

When considering an accident as a symptom of system dysfunctioning, the analyst is led to search for other symptoms, particularly symptoms of the same dysfunctioning which do not constitute accidents. Incidents, breakdowns and near-accidents belong to this latter category. A decisive issue for safety is to know to what extent such symptoms can give clues to the same dysfunctions as accidents. The question in this case is the determination of the relationship between reliability and safety. Does the former embrace the latter? Does any action which favors reliability necessarily favor safety? This question was examined in detail elsewhere (Leplat, 1982); consequently, I shall just mention some of its essential aspects. Three cases can be mentioned among the possible relationships between dysfunctions, incident and accident. In the first case, incidents are only the symptoms of dysfunctioning. A starter which does not work properly (dysfunctions) is the source of possible incidents (repeated attempts at starting the engine, which exhausts the battery) but does not result in an accident.

In the second case, the dysfunctioning generates incidents which themselves create accidents. There are several examples in the literature showing that accidents occurred at the end of a series of incidents (Faverge, 1967). The mechanisms through which incidents result in accidents are diverse. An accident may be a direct consequent of an incident (for instance a trolley leaves the track and a worker is thrown down) or an indirect one. The latter category includes accidents which occur when attempting to counteract an incident (for instance, a worker gets hurt when putting the trolley back on its track).

Finally, the same dysfunctioning may result both in incidents and accidents which are not related to each other. For instance, in some cases, the fact that the sling of a crane is not properly hooked (dysfunctioning) may result in some cases in material damages (damaged objects), in other cases in physical damage to the workers on the ground.

Simple schematic appreciations of this type will often have to be made more complex in order to take into account observed accidents. The same dysfunctioning may in fact result in the three different cases just mentioned.

In practice, the important point is to establish a link between incident and accident in order to use incidents — which are always more frequent — as predictors of accidents and to reveal dysfunctionings. The most systematic studies undertaken within this perspective concern road safety and aim at evaluating the danger of certain road configurations (intersections particularly), as well as providing us with elements of diagnosis of these situations.

Incidents or near-accidents considered here are called traffic conflicts. The definition of this conflict is difficult. Malaterre et al. (1976) defined it

as follows: "There is a conflict when at least a utilizer is obliged to undertake an avoidance action, such as braking, or accelerating or changing lane or direction in order to avoid a probable accident" (p. 8). Identifying as well as evaluating a conflict raises difficult problems which cannot be dealt with here. I shall just report some characteristic results of these studies.
(1) The correlation between the number of conflicts and the number of accidents varies with the severity of the conflict: it increases as more severe conflicts are taken into consideration (Malaterre et al., 1978; Russam et al., 1972).
(2) The relationships between conflicts and accidents vary with the types of utilizers, the types of conflicts (Table I) and the types of intervention.
(3) In some cases, accidents and conflicts seem to belong to different types. "These cases are observed on some difficult sites where the dangerous character of the cross-road is not due to the frequency of conflictual manoeuvres, but to a very particular dysfunctioning which is difficult to discover, which becomes evident in certain conditions and in an environment which is generally already structured to avoid conflicts" (Malaterre et al., 1979, p. 18).

TABLE I

Relationships between conflicts and accidents as function of the type of utilizers and conflicts (1, 2, 3). The arrows represent the direction of the vehicles involved in the conflict
(after Malaterre and Muhlrad, 1978)

	Light vehicles	Heavy lorries	2 wheels	Pedestrians	1 →→	2 →↑	3 →↙
% conflicts	54	10	27	9	19	27	11
% accidents	23	3	68	6	8	44	22

These examples all point to the fact that if incidents may provide us with clues to safety which are sometimes very useful, it is always necessary to examine very carefully their real relationships with accidents, by using either clinical studies or statistical studies. An example of the exploitation of incidents (that the author called "non-injury accidents") in the steel industry was recently given by Laitinen (1982).

ACCIDENTS AND CHANGES IN THE SYSTEM

Change and deviation are important concepts both in systems theory and in accident theories. Kjellén (1983) pointed to the usefulness and scope of these notions in a more detailed way than I shall. I shall just underline some aspects of the relationships between accident and change.
In his book on cybernetics, Ashby (1970) emphasizes the importance of

the concept of change. This notion of change is connected with the notion of variance. Johnson (1980), who devoted many studies to accidents, entitled a chapter of his synthesis "Change is the mother of trouble" and wrote "if a system has been operating in a stable manner but now has troubles, change is the cause of the problem" (p. 53) and, "it is intuitively obvious that if tasks and jobs comparable to those involved in an accident have been conducted in the past without incident, changes and differences provide a logical focal point in accident investigation" (p. 57). Change is conceived as the destabilizing element in a system which has no adequate response for it.

However, the notion of change which apparently seems clear is not easy to introduce into the investigations. The first essential questions that it raises is to determine in comparison to what reference event a given event is considered as a change. Two cases are most usually considered as constituting a change:

- Change in comparison with what is officially prescribed or planned.
- Change in comparison with the usual situation, such as is habitual in the place where it occurs.

In the first case, change is considered as a deviation in comparison with the norm represented by the prescriptions of the organization, these prescriptions being meant to minimize the risk. The reference is to the rationality of the organization.

In the second case, a change involves the idea of destabilization. The usual situation has acquired some stability and a deviation from this stability invalidates to different extents the typical processes of functioning of the stable situation and/or triggers others which are not necessarily adapted to change.

Other cases are conceivable, for example a change in comparison with an analogous situation which occurred immediately before, or earlier, or with a situation which shares some resemblance with the present one (for instance the same work place in a different workshop or a different plant), etc. One may also wonder who is experiencing the change. Is it the expert? Is it the

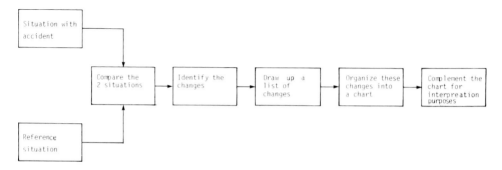

Fig. 2. Diagram of the different stages in the search for and exploitation of changes.

subject? The INRS method (for example, Cuny 1977) to which I shall refer, considers changes which occur in comparison with the habitual situation (Fig. 2). The problem is to define what is habitual. Researchers who devised this method proposed as a criterion of normality the condition which is considered as habitual by operators or analysts who are very familiar with the work place.

This definition is not very clear but, in the analysis of the genesis of an accident, it can be shown that errors in the definition of normality are not very grave.

Different classes of changes have been put forward. Johnson (1980) mentioned some of them. The INRS advanced a sociotechnical classification in terms of individual, task, equipment and environment. I proposed a classification which combines the distinction between action and state with different types of agents (individual, materials, machine, environment, organization). Specifying the latter categories may be very useful when confronted with homogeneous situations (in other words specifying the type of materials, machine, etc). Cuny proposed also to distinguish intra-element variations (e.g., the degradation of a machine) from interelement variation (e.g., the removal of a machine).

The changes involved in an accident can be organized into a diagram (as in the INRS method). This diagram of changes is only a part of the diagram of causes and it is often useful to complement it by mentioning some causes which permit a better understanding of the genesis of an accident. Then, the organization of changes constitutes only the first stage in the analysis of accident occurrences. They serve as guides and can also suggest useful preventive action. With Rasmussen (1983), we recently proposed a procedure for exploiting these diagrams systematically. The study of several of these diagrams can also sometimes permit the identification of typical configurations that appear in different charts and represent categories of dysfunction-

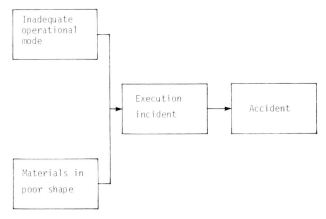

Fig. 3. An example of a typical configuration of a dysfunctioning likely to lead to an accident (after Meric and Szekely, 1980).

ing. Figure 3 presents one example among the different cases mentioned by Meric et al. (1980). The labels in the boxes constitute the general title of a class of particular elements characterizing the collected diagrams of accidents.

SYSTEMS OF REFERENCE OF ACCIDENTS IN INDIVIDUAL APPRECIATION

The systems within which an accident may be inserted are numerous. An accident may be attributed to a dysfunctioning at the organization level, or at the sociotechnical level constituted by the workshop of the workteam and their technical environment, or at the level of the elementary man—machine system. But an accident may also be related to the history of the individual, his personality, his abilities, etc., which also constitute systems interacting with the former. The relative importance of each of these systems varies from one accident to another. A historical review of studies on accidents reveals that the systems of reference which are preferred vary with the period. For a long time, psychologists considered individual characteristics as the main determinant of accidents, and the question of a predisposition to accidents was widely discussed. Later on, different systems appeared. If we now consider no longer the history but the different persons and groups in our society, we may observe an analogous phenomenon, namely that although each person's appreciation of the causes of an accident is different, it does not follow that they are completely wrong. A not-very-recent inquiry (Vibert, 1957) on modernization, involving 310 French workers and 7 industries asked the following question: "In your opinion, what is the cause of an accident? Workers' lack of attention, too fast a work-pace, disregard of safety instructions, hazard, bad functioning of machines, fatigue, or lack of work coordination within a team". It appeared that "workers satisfied with their job, integrated into and participating in the enterprise, attribute accidents mainly to personal causes" (workers' lack of attention, disregard of safety instructions). Contrarily, workers who are not very satisfied, with a low degree of integration and participation invoke more often non-personal causes that imply that the enterprise is responsible. This inquiry revealed a relationship between the attitude towards the enterprise and workers' appreciation of the causes of accidents.

More recently, Sundbo (1980) showed differences in the attribution of the causes of accidents with the identity of the person questioned: victims, safety executives, management.

A systematic study of causal attribution was conducted in our laboratory by Kouabenan (1982). 320 workers from a public factory were asked questions analogous to the questions asked in the above-mentioned inquiry. Furthermore, workers were requested to interpret reports of accidents. Kouabenan distinguished three groups of subjects, depending on their qualification. The results show that the lower the position in the hierarchy, the greater the tendency to attribute the causes of accidents to factors linked to work organization (inadequacy or lack of safety measures, faulty materials, bad working conditions, etc.); individuals who have a high position in the hierarchy tend to attribute responsibility for accidents to the workers (lack of attention, of caution, inexperience, disregard of safety rules). It seems that accidents are attributed to the factors in which individuals are less directly involved. In his conclusion, the author underlines that the causal attribution of an accident seems to depend on some characteristics of the victim and of the analyst (hierarchical status, degree of involvement, satisfaction), as well as on the relationships between the victim and the analyst, and on the severity of the accident. This study was elaborated and analyzed within the framework of the attribution theory (Kelley, 1972). The same theory inspired several experimental studies which aimed at a better specification of the factors and mechanisms of attribution.

When dealing with safety problems, the variation in the systems preferred by individuals when making their analysis should be taken into account (a) In order to avoid bias which might distort or impoverish the analysis, it is necessary that persons having diverse professional status and different training participate in the analysis. In this way, there will be more opportunity to take into account the whole set of factors affecting the genesis of an accident.

(b) It is possible — but experiments should investigate this point to confirm it — that operators' behavior depends partly on their appreciation of the work situation and that, by modifying this appreciation through adequate training, one will contribute to the development of safety behavior. It was shown in some studies that the attribution of accidents to hazard was more frequent in individuals having a low degree of training; it would be interesting to see if the same individuals are also less cautious. It would also be useful to have the means to evaluate the influence upon safety behavior of training in the analysis of accidents, particularly of training in an analysis showing the role of the different intervening systems.

CONCLUSIONS

I have attempted to present some orientations suggested by the "systems-thinking" approach in the study of accidents. I shall now try to extract the main general characteristics of this approach. At first, it seems that the systems theory in the analysis of accidents does not diverge fundamentally from other theories advanced. All theories extract and organize certain variables that may be considered as defining a system. In my view, systems theory is an invitation to coordinate the various theories of accidents which are always partial theories. An accident is a phenomenon resulting from the intervention of a set of variables for which there is no simple model. Their study will often necessitate the combination of several models, and the concepts of the systems theory may help towards this combination and organization.

However, systems theory does not specify the variables to be taken into account, and the components of the system that will be investigated, as well as its boundaries and functioning, remain to be defined. For this, partial models which are already well known may be useful.

The systems theory approach underlines the limitations of all interpretations in terms of a unique cause. Accidents do not have one single cause but rather a network of causes. The problem is not only to identify these causes, to locate them in terms of their distance from an accident (on a proximal/distal axis) but also to evaluate their respective importance. Multiple causes are always present, quite often very intricate, but their role is not equivalent either in a given accident or in a set of accidents.

One should also note that the systems approach emphasizes the problems of functioning and production. To think of an accident in terms of a system is to search for the mechanisms which produced it and for the characteristics of the system which may give an account of this process.

Finally, it should be kept in mind that systems explaining occupational accidents involve either one or several individuals or the variables which characterize them. In itself this characteristic is the source of great complexity, not only because of the interindividual variability, but also because individuals and groups may adopt their own goals which are not necessarily the same for all and which do not necessarily coincide with the goal of the enterprise. Men's activities in their work are oriented and directed by the appreciation they have of personal, technical or organizational systems which characterize them or into which they are inserted. This appreciation is quite often more difficult to apprehend and to control than their objective itself.

I hope I have been able to outline several orientations which may enable an approach to be made to this complexity and facilitate the analysis. Unfortunately, however, there is no simple recipe for doing so.

REFERENCES

Ashby, W.R., 1970. An Introduction to Cybernetics. Chapman and Hall, London, 295 pp.
Cuny, X., 1977. An accident analysis method. In: Research on occupational accident, pp. 37—41. Swedish Work Environment Fund, Stockholm.
Emery, F.E., 1969. Systems Thinking. Penguin Books, Harmondsworth, 398 pp.
Emery, F.E. and Trist, E.L., 1960. Socio-technical systems. In: F.E. Emery (Editor), Systems Thinking. Penguin Books, Harmondsworth (1969), pp. 281—296.
Faverge, J.M., 1967. Psychosociologie des Accidents du Travail. P.U.F., Paris, 160 pp.
Faverge, J.M., Houyoux, A., Olivier, M., Querton, A., Laporta, J., Poncin, A. and Salengros, P., 1970. L'Organisation Vivante. Editions de l'Institut de Sociologie, Bruxelles, 198 pp.
Hackman, J.R., 1969. Toward understanding the role of tasks in behavioral research. Acta Psychol., 31: 97—128.
Herbst, P.G., 1974. Socio-Technical Design. Tavistock Publications, London, 242 pp.
Johnson, W.G., 1980. MORT Safety Assurance Systems. M. Decker Inc., New York, 525 pp.
Kelley, H.H., 1972. Causal schemata and the attribution process. In: E.A. Jones et al., Attribution: Perceiving the Causes of Behavior. General Learning Press, Morristown, N.J., pp. 151—174.

Kjellén, U., 1983. The deviation concept in occupational accident control — theory and method. Royal Institute of Technology, Report No.Trita-AOG-0019, Stockholm.

Kouabenan, D.R., 1982. Représentation de la genèse des accidents du travail: déterminants des attributions causales. Thèse de 3ème cycle, E.P.H.E., Paris V, 345 pp.

Laitinen, H., 1982. Reporting non-injury accidents: a tool in accident prevention. J. Occupational Accidents 4 (2—4): 275—280.

Leplat, J., 1982. Fiabilité et sécurité. Le Travail Humain, 45 (1): 101—108.

Leplat, J. and Cuny, X., 1979. Les Accidents du Travail. Coll. "Que Sais-Je" No. 1591, P.U.F., Paris, 126 pp.

Leplat, J. and Hoc, J.M., 1983. Tâche et activité dans l'analyse psychologique des situations. Cahiers de Psychologie Cognitive, 3 (1): 49—64.

Leplat, J. and Rasmussen, J., 1983. Analysis of human errors in industrial incidents and accidents for improvement of work safety. Report to be published.

Malaterre, G. and Muhlrad, 1976. Intérêt et limite du concept de conflict de trafic et quasi-accident dans les études de sécurité. Document interne, Monthlery, ONSER, 37 pp.

Malaterre, G. and Muhlrad, 1978. Mise au point d'une méthodologie des conflicts de trafic. Document interne, 79.41.034, Monthléry, ONSER, 39 pp.

Meric, M. and Szekely, J., 1980. Diagnostic de sécurité préalable à la définition d'actions de prévention. Rapport No. 399/RE et annexe I.N.R.S., Vandoeuvre.

Russam, K. and Sabey, B.E., 1972. Accidents and traffic conflicts at junctions. Report L.R. 146, Road Research Laboratory, Crowthorne.

Sundbo, J., 1980. Tilskadekomoster — forekomst of risitofaktorer. I Kommission hos teknisk forlag, København.

Vibert, P., 1957. La représentation des causes d'accidents du travail. Bull. du CERP, VI (4): 423—428.

X, 1969. Recherche communautaire sur la sécurité dans les mines et la sidérurgie A — Sidérurgie. Etude de physiologie et de psychologie du travail No. 4, CECA, Service des publications des communautés européennes, Luxembourg.

ACCIDENTS, AND DISTURBANCES IN THE FLOW OF INFORMATION

JORMA SAARI

Institute of Occupational Health, Haartmannibatu 1, SF-00290 Helsinki (Finland)

ABSTRACT

Saari, J., 1984. Accidents and disturbances in the flow of information. *Journal of Occupational Accidents*, 6: 91—105.

Man continuously interacts with his environment. The undisturbed exchange of information between man and his environment can be assumed to be an important prerequisite to accident avoidance. Man should be aware of possible sources of harmful energy in the environment. He also should be able to assimilate the information about the sources of energy for the choice of correct action. The relevance of this model was tested with two samples of accidents and by comparisons. The samples were taken from the light metal industry and the printing industry. It was found that accidents in the light metal industry occurred in environments where the size of dangerous areas was less than average, whereas the opposite was true in the printing industry. The possible explanations are discussed, and the overall applicability of the model is evaluated.

INTRODUCTION

The purpose of this paper is to evaluate the validity of energy flow models in the selection of preventive measures for accidents. It has been claimed that disturbances in the exchange of information between man and the environment have such a great effect on the occurrence of accidents that such disturbances must be considered when preventive measures are chosen. The paper describes some of the author's results and briefly reviews other literature regarding the validity of these two approaches.

An injury is the unwelcome consequence of an accident. A necessary, but not the sole, requisite for an injury is contact with a harmful substance or energy, e.g., mechanical force, chemical substance, thermal or electrical energy, radiation, etc. An injury may also be caused by a critical lack of energy (e.g., lack of oxygen in the environment).

The criticality of an accident (the immediate consequences, see Goeller, 1969) is a function of the intensity of the energy and the tolerance of the human body, which depends on the way the energy and body come into contact (which part(s) of the body, the type and amount of energy). The criticality is both deterministic and stochastic in nature.

The source of injurious energy is either external or internal to man.

Moving machine parts or a poisonous gas in the air are examples of external sources. Examples of internal sources are energies due to walking, work movements and even standing, which energies are transformed into impact against the floor, the sharp edge of a machine, etc. as a consequence of slipping, stumbling or other uncontrolled and unintentional movements.

Thus, the areas of a working space where the sources of harmful energy exist can be defined as danger zones. The boundaries of the danger zones vary with (a) the type of energy, (b) the part of the body possible affected and its tolerance, and (c) the phase of the work process. Figure 1 gives a simplified illustration of danger zones. An injury occurs if man comes into contact with a danger zone.

The critical question for the prevention of accidents is the influence of the volume of the danger zone on the probability of accidents. Several hypotheses have been put forward in the literature. Perhaps the most common assumption is that of linearity (especially in textbooks of safety engineering). The probability of injuries would be a linear function of the volume of the danger zone (Fig. 2), or at least the probability of accidents always decreases as a consequence of the reduction of the danger zone (Skiba, 1973; Johnson, 1980; Haddon, 1980). According to this hypothesis, preventive measures of types (a) the elimination of one or several of the danger zones, (b) putting a barrier between man and the danger zone, and (c) the reduction of the exposure time to the danger zone, would always lead to fewer accidents.

Another hypothesis might be that tasks performed determine the worker's distance from the danger zones and thus modify the probability of accidents. Accident rate is known to vary with tasks (Saari, 1976; Saari and Lahtela, 1981).

The third hypothesis is that personality factors or the individual's other

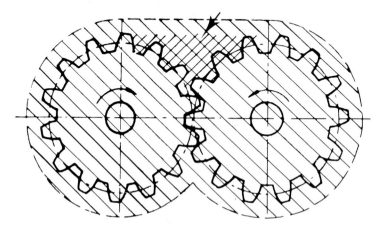

Fig. 1. An illustration of the danger zone concept. The gear transmission creates two danger zones: a) the vicinity of rotating gears, and b) the nip area. The size of these danger zones depend on the rotating speed, the size of gears, the moment, etc.

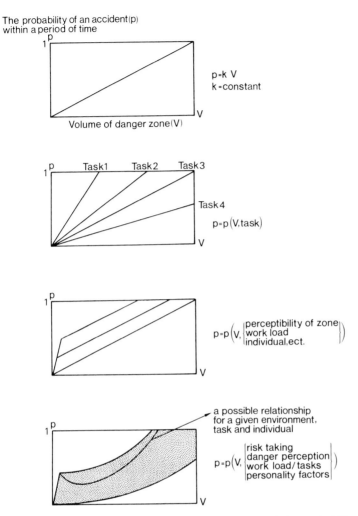

Fig. 2. Four hypotheses regarding the relationship of the probability of accidents and the volume of the danger zone.

characteristics increase the probability rapidly when the danger zone exceeds zero. According to the accident liability theory, each individual has a different liability to accidents, though this also depends on the environmental characteristics (Hale and Hale, 1971). The perceptibility of the zone, work load, etc., could have the same effect (Zohar, 1978; Zimolong, 1979). These factors might have complex interactions.

Brown et al. (1969) found that the introduction of a task concurrently with driving may lead to interferences between the two tasks and as a consequence relax criteria used in decision-making, or impair perception due to continuous switching of sensory modes.

The fourth hypothesis suggests that man, environment and machine form a system in which a change in one of these components may induce a change in another (Saari, 1982).

As a consequence of the reduction of a danger zone, man may behave in a new manner and compensate for the improvement by being more indifferent to dangers. For example, a new strategy in decision-making may be adopted: (a) being indifferent to safety distances, (b) changing active search and the processing of information, (c) neglecting precautionary actions against hazards (Hoyos, 1976).

The compensation theory has been discussed especially in connection with traffic safety. It has been suggested that a driver attempts to maintain a constant risk rate per minute (Wilde, 1976). O'Neil (1977) has shown mathematically that improvements in objective safety (e.g., better roads) may, with certain assumptions, lead to a higher frequency of accidents. According to empirical observations drivers used the increased objective safety created by studded tyres partly for improved safety and partly for improved performance (Rumar et al., 1976). Smith and Lovegrove (1983) found some indications of compensation at intersections with a stop sign.

A basic question for selecting effective preventive measures against accidents is whether the reduction of the volume of danger zones leads to improved safety or not. The traditional preventive measures of safety engineering (elimination of sources of harmful energy, machine guards, etc.) are based on this assumption.

The aim of the following data is to assess whether lower accident rates are related to smaller danger zones or not. A positive answer should lead to the following findings:
- the danger zones in industry of low accident frequency should be smaller than in industry of high accident frequency
- the frequencies of specific accident types are related to the volume of the respective types of danger zone (e.g., many unguarded moving machine parts should lead to many injuries being caused by them)

MATERIAL* AND METHOD

The material was collected from seven companies operating in the light metal industry and ten companies in the printing industry. These two industries were selected because they were known consistently to differ in accident frequencies. The total number of employees was 1584 and 2179, respectively (Table 1). Each accident leading to at least one day's absence was studied over a six month period. The number of accidents was 130 in the light metal industry and 54 in the printing industry. Accident rates were 136 accidents/million man-hours and 45 accidents/million man-hours.

---

*The material was collected in cooperation with Mr. J. Lahtela, who also helped in the analysis of data. The present author is responsible for the treatment of data and for the conclusions in this paper.

TABLE 1

The material of the study

	Light metal industry	Printing industry
Companies	7	10
Workers	1584	2197
Accident frequency (accidents/10^6 man-hours)	136	45
Accidents	130	54
Controls	222	300

In the same companies, every sixth work place was studied, using the same method as in work places where accidents occurred. These controls acted as comparison groups for accident groups, thus giving the average distributions of the variables in the companies studied. The number of work places in the control group was 223 in the light metal industry and 300 in the printing industry.

Each work place was observed, and each worker interviewed, using special prepared forms. The description of the danger zones included the following variables:
(a) the type of energy,
(b) the source of energy,
(c) the volume of the danger zone,
(d) the movement of the danger zone, and
(e) the proportion of the time in which the zone existed.

RESULTS

Table 2 presents the sources of energy in the accidents studied. The different sources of energy are classified into external and internal to man. Roughly half the injuries were caused by external sources in both industries. Otherwise the distributions differ considerably. The most frequent external source of energy was moving machine parts in the printing industry and falling objects in the light metal industry.

Regarding the internal sources of energy, there is one quite clear difference. Uncontrolled movements of upper extremities resulting in an impact against a sharp edge or other surface are more frequent in the light metal industry than in the printing industry.

Table 3 presents the incidence rates of injuries caused by different sources of energy. The exposure time is an estimate of time that workers really were exposed to different types of energies indicated in the Table. Thus the exposure time is not the total working time but a much more correct estimate of the real exposure. The exposure time was calculated from control groups. The criterion for being exposed to a danger was that at least one source of

TABLE 2

The source of injuring energy

Source of energy	Light metal industry		Printing industry	
	N	(%)	N	(%)
I. External to man	61	47	27	50
Moving machine parts	8	6	17	32
Moving objects	38	29	6	11
• falling objects	25	19	5	9
• flying particles	13	10	1	2
Overexertion	15	12	4	7
II. Internal to man	64	49	22	41
Uncontrolled whole-body				
movements	21	16	11	20
• slipping	11	8	7	13
• stumbling	7	5	1	2
• stepping on	2	2	2	4
• falling	1	1	1	2
Uncontrolled movements of the				
upper extremities	43	33	11	20
III. Others				
(combination, unclassified)	5	4	5	9

TABLE 3

The incidence rates of injuries caused by different sources of energy (exposure data from the control work places)

Source of energy	Incidence rate (accidents/10^2 man-year's exposure)	
	Light metal industry	Printing industry
I. External to man		
Moving parts of machines	2	6
Moving objects		
• falling objects	3	7
• flying particles	4	9
II. Internal to man		
Walking	21	5

harmful energy was at such a distance that contact was possible between worker and the source. The criterion (distance) depended thus on the type of energy.

The incidence rates of injuries caused by external energies are higher in the printing industry than in the light metal industry. The result is an indication that the probability of injuries varies despite similar exposure to harmful energies.

The danger zone due to moving parts of machines is a little larger in the light metal than in the printing industry. Table 4 presents more detailed results concerning these danger zones. The number of injuries caused by moving parts of machines was higher in the printing industry (Table 2).

The proportion of time when the danger zones existed was about the same in both industries. However, in the accident work places in the metal industry the average activation time of the danger zones was longer than in the control work places. No such difference existed in the printing industry.

Regarding falling objects, the danger zone was much larger in the light metal industry than in the printing industry. This explains the incidence rates (Table 3). Table 5 presents more detailed results from moving objects.

TABLE 4

Percentage of work places at which there existed 0.1 or >1 danger zone owing to moving parts of machines. The boundary of the mobile danger zone moves while the boundary of the fixed zones remains in the same place

Number of danger zones/work place	Parts of machines					
	Rotating parts		Reciprocating parts		Nips	
	Fixed	Mobile	Fixed	Mobile	Fixed	Mobile
Light metal industry						
Accident group						
0	93	91	80	93	100	98
1	5	8	19	7	0	1
>1	2	1	1	0	0	2
Control group						
0	74	86	76	92	97	98
1	13	10	19	8	2	1
>1	13	4	5	0	1	1
Printing industry						
Accident group						
0	81	100	83	96	90	91
1	2	0	13	4	5	7
>1	17	0	4	0	5	2
Control group						
0	86	98	89	100	94	99
1	6	1	7	0	3	1
>1	8	1	4	0	3	0

TABLE 5

Percentage of work places at which there existed 0.1 or >1 danger zone owing to falling object and flying particles (for the definition of fixed and mobile see Table 4)

Number of danger zones/work place	Falling objects (>10 Nm)		Flying particles	
	Fixed	Mobile	Dust (harmful only to eyes)	Scarf and other larger fragments
Light metal industry				
Accident group				
0	95	45	81	98
1	3	42	18	2
>1	2	12	1	0
Control group				
0	43	27	76	83
1	16	16	24	17
>1	42	57	0	1
Printing industry				
Accident group				
0	100	80	100	100
1	0	18	0	0
>1	0	2	0	0
Control group				
0	99	93	99	99
1	1	5	1	1
>1	0	2	1	0

Again the result is an indication that with similar exposure to moving objects the probability of accidents varies.

For injuries caused by the energy of walking, the incidence rate is higher in the light metal industry than in the printing industry. Table 6 presents results concerning the volume of the danger zone of walking.

The result again is an indication in the same direction, that with the same exposure the probability may vary. This is an even stronger indication than the previous ones, as walking must be a similar activity in both industries while being exposed to external energies in work activities is more varied.

A possible explanation of the result is that walking is a more demanding task in the light metal industry than in the printing industry because of more difficult environmental conditions (Table 7). Objects, rubbish, oil, etc. on the floor are far more common in the metal industry, thus making it difficult to walk without allocating some perceptual capacity to the walking surface. If other simultaneous demands exist, the possibility of exceeding the capacity of information handling becomes real.

TABLE 6

The proportion of time spent walking as a percentage distribution at work places studied (%)

Time	Accident group	Control group
Light metal industry		
None	8	6
<10%	21	69
11—50%	68	22
51—90%	4	4
>90%	—	—
Total time of exposure in man-years*	190	100
Printing industry		
None	—	16
<10%	9	26
11—50%	87	52
51—90%	4	5
>90%	—	1
Total time of exposure in man-years*	320	230

*During the study period.

TABLE 7

Some environmental characteristics related to moving at a workplace (%)

Number of characteristics observed/work place	Objects on the floor		Rubbish on the floor	Oil, etc., on the floor
	Fixed	Mobile		
Light metal industry				
Accident group				
0	65	60	85	97
1	24	15	}15	}3
>1	11	25		
Control group				
0	26	22	12	13
1	24	14	}88	}87
>1	50	54		
Printing industry				
Accident group				
0	54	76	87	93
1	19	11	}13	}7
>1	28	13		
Control group				
0	69	77	90	97
1	15	15	}10	}3
>1	17	9		

CONCLUSIONS

The determination of the real volume of danger zones is not an easy task. In this study a dichotomized scale was used. It creates some errors in the results. The volumes were estimated during observation at the work places. On this basis the simplified analysis does not produce so large an error that it would violate the conclusions.

The results are clearly an indication suggesting that the probability of accidents does not always decrease by reducing the volumes of danger zones. The study was comparative by design and cannot therefore finally prove this conclusion. But the study proves that the incidence of accidents is dependent also on other factors — not only on the volume of danger zones.

The incidence rates for similar danger zones vary largely. Danger zones created by moving parts of machines are slightly smaller in the printing industry, but the number of accidents was much higher than in the metal industry. Danger zones due to falling objects are much larger in the metal industry but the incidence rate of accidents is again higher in the printing industry.

The mechanisms leading to this result are most probably different. A possible explanation for the frequent occurrence of injuries caused by machine parts in the printing industry is the demanding type of tasks. Typically they are service and maintenance tasks which require continuous planning and continuous observation of the environment. In the metal industry the tasks performed in the vicinity of machine parts are less varied or even repetitive production tasks.

For falling objects the explanation might be different. These form such rare danger zones in the printing industry that recognising and being aware of falling objects becomes difficult. People do not expect objects to fall and do not devote their attention to this possibility.

Walking was more common in the printing industry but injuries caused by energy originating from walking were less common than in the light metal industry. Thus the mere existence of energy sources is not a sufficient predictor of accidents. The possible explanation is that walking areas in the metal industry are worse. Either there is no awareness of deficiencies of floors or the motor performance fails.

Altogether, the results suggest that, in addition to the existence of danger zones, other factors contribute considerably to the occurrence of accidents. Disturbances in the exchange of information between man and his environment are one possible factor.

Information flow as a determinant of injuries

The processing of information is in any task an essential factor in coping with a situation. A natural assumption thus is that a disturbance in the information flow between man and the environment should affect coping with danger zones. Hence, disturbances in the processing of information would explain the curvilinearity of the relationship between danger zones and accident probability.

Thus, the occurrence of accidents would depend on how well man is aware of the danger zones and can avoid entering them. This approach has been proposed and applied in two ways:

(a) by preparing an analytical model for information processing during work (e.g., Goeller, 1969; Hale and Hale, 1970; Wigglesworth, 1972; Smillie and Ayoub, 1976; Fell, 1976; Corlett and Gilbank, 1978; Häkkinen, 1979; Feggetter, 1983; Hoyos et al., 1981; WHO, 1982) or

(b) by analysing the information flow in a situation requiring evasive action (Surry, 1968; Lawrence, 1974; Andersson et al., 1976).

A number of researchers have used the information flow model when explaining the observations of accidents or of accident-like occurrences. Some of the earlier work was carried out in traffic situations (Russel Davis, 1958, 1966; Buck, 1963; Goeller, 1969).

By analysing individual accidents, Kano (1975) noticed that an accident may occur in relatively easy or habitual activities when the injured person was unaware of the danger zone.

The cognitive processes in relation to an external object are determined by its functional relationship with the real task. The chances of neglecting a part of the environment are increased if the real task has no functional relationship with the danger zones (Kano, 1975).

Hoyos et al. (1981) found that accidents were related to high demands in information processing (a) in avoidance of dangers while performing work or (b) when the degrees of freedom in work performance are many and wide. These results correspond closely with Saari's and Lahtela's results (1979, 1981).

Work is a dual task. On the one hand the danger zones have to be monitored as a secondary task, while performing the productive work is the primary

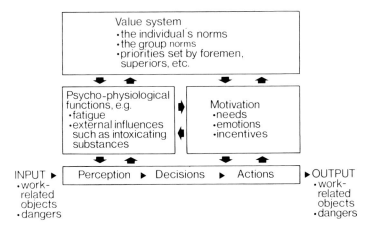

Fig. 3. The flow of information is dependent on several internal and external factors.

task. These may conflict, as already mentioned (Brown et al., 1969; Kano, 1975). The extra demands on processing were demonstrated experimentally by Zohar (1978).

The information flow is influenced by the suitability of information for perception, comprehension and decision-making. Past experience, as well as instruction received, affect orientation towards various sources of information. However, the process is influenced by several internal factors, such as the state of the organism (fatigue, etc.), motivation and values assumed (Fig. 3). All these make the whole matter very complex and the mechanisms possibly leading to a disturbance are therefore many.

Figure 4 summarizes failures in information handling that in our studies have most often been found to contribute to the occurrence of accidents (Saari, 1976; Saari and Lahtela, 1979, 1981).

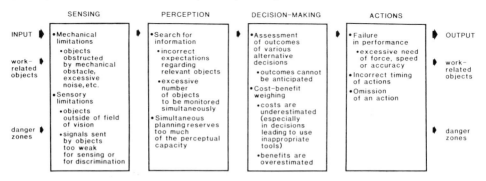

Fig. 4. Most common mechanisms leading to error in information processing according to the author's previous studies (Saari, 1976; Saari and Lahtela, 1979a, b; Saari and Lahtela, 1981).

The validity and practicability of the information flow model

The information flow model is an overall model of accident occurrence. Failures and disturbances may occur in the flow of information in many ways. Therefore, it may be impossible to validate the model. However, some benefits, problems and limitations can be listed.

Some of the benefits are:

- the model is independent of accident definition based on the consequences of the accident process (= injuries)
- it gives deeper insight into the causal mechanisms, and more effective and relevant countermeasures (Hale and Hale, 1970; Wigglesworth, 1972; Fell, 1976; Corlett and Gilbank, 1978)
- it contains a systems approach, thus allowing for the observation of the interactive influences of various factors
- it improves the reporting of incidents (Fell, 1976) and their description (Hale and Hale, 1970)

- it accords with the traditional techniques of industrial safety engineering (Wigglesworth, 1972)
- the model may give an understandable and useful framework for planning, as well as analysis.

Some of the drawbacks are:

- the model is not primarily concerned with causal factors but with the mechanisms through which they operate (Wigglesworth, 1972)
- by emphasising the role of human error it may appear concerned with blame, since the term "error" commonly carries a connotation of fault (Wigglesworth, 1972)
- there still exists a lack of valid models explaining human behavior in relationship to environmental conditions (Zimolong, 1976)
- the modes and mechanisms of disturbances are manifold, and therefore more accurate models of isolated situations are required
- pre-accident analysis is more inaccurate than post-accident analysis (Hale and Hale, 1970).

Some of the limitations are:

- the model does not sufficiently take into consideration influences and determinants other than man—machine interfaces (or more generally man—environment interfaces); social influences and the values of individuals or groups are not incorporated (Fig. 3)
- the model does not apply to purely technical events and failures
- it is not applicable to the analyses of the processes in the formation of working conditions; it takes the working environment as it stands and may be used for analysing the given man—machine system.

Despite the problems and limitations, the information flow model is worth elaborating. The traditional straightforward safety engineering techniques aiming at the reduction of danger zones need to be replaced with models differentiating the effectiveness of preventive measures to a greater extent than at present.

REFERENCES

Andersson, R., Johansson, B., Lidén, K., Svanström, K. and Svanström, L., 1976. Accidents at work — a study conducted at Malmö. Occupational Accident Research, The Swedish Environment Fund, Stockholm, pp. 171—180.

Anon., 1979. Definition of danger zones and evaluation of danger grading in connection with band saws. IVSS-Bulletin, pp. 3—34.

Brown, I.D., Tickner, A.H. and Simmonds, D.C.V., 1969. Interference between concurrent tasks of driving and telephoning. Journal of Applied Psychology 53: 419—424.

Buck, L., 1963. Errors in the perception of railway signals. Ergonomics 6: 181—192.

Corlett, E.N. and Gilbank, G., 1978. A systematic technique for accident analysis. Journal of Occupational Accidents 2: 25—38.

Feggetter, A.J., 1983. A method for investigating human factor aspects of aircraft accidents and incidents. Ergonomics 25: 1065—1075.
Fell, J.C., 1976. A motor vehicle accident causal system: the human element. Human Factors 18: 85—94.
Goeller, B.F., 1969. Modelling the traffic-safety system. Accident Analysis and Prevention 1: 167—204.
Haddon, W., 1980. The basic strategies for reducing damage from hazards of all kinds. Hazard Prevention 16: 8—12.
Hale, A.R. and Hale, M., 1971. A review of the industrial accident research. Her Majesty's Stationary Office, London, p. 96.
Hale, A.R. and Hale, M., 1970. Accidents in perspective. Occupational Psychology 44: 115—121.
Hoyos, C. Graf., 1976. Entscheidungsautonomie in Mensch—Maschine-Systemen. Zeitschrift der Arbeitswissenschaften 30: 216—220.
Hoyos, C. Graf., Gockeln, R. and Palecek, H., 1981. Handlungsorientierte Gefährdungsanalysen an Unfallschwerpunkten der Stahlindustrie. Zeitschrift der Arbeitswissenschaften 3: 146—149.
Häkkinen, S., 1979. Accident theories. Acta Psychologica Fennica VI: 19—28.
Johnson, W.G., 1980. MORT-safety assurance systems. Marcel Dekker Inc., New York, p. 525.
Kano, H., 1975. The genesis and mechanism of human error in accidents resulting in latent danger. Journal of Human Ergology 4: 53—63.
Lawrence, A.C., 1974. Human error as a cause of accidents in gold mining. Journal of Safety Research 6: 78—88.
O'Neill, B., 1977. A decision-theory model of danger compensation. Accident Analysis and Prevention 9: 157—165.
Rumar, K., Berggrund, U., Jernberg, P. and Ytterbom, U., 1976. Driver reaction to a technical safety measure — studded tires. Human Factors 18: 443—454.
Russel Davis, D., 1958. Human errors and transport accidents. Ergonomics 2: 24—33.
Russel Davis, D., 1966. Railway signals passed at danger: the driver's circumstances and psychological processes. Ergonomics 9: 211—222.
Saari, J., 1976. Technical prevention of industrial accidents. Occupational Accident Research, The Swedish Work Environment Fund, Stockholm, pp. 123—131.
Saari, J. and Lahtela, J., 1979a. Characteristics of jobs in high and low accident frequency companies in the light metal working industry. Accident Analysis and Prevention 11: 51—60.
Saari, J.T. and Lahtela, J., 1979b. Work structuring and safety. In: Sell, R.G. and Shipley, P., Satisfactions in Work Design: Ergonomics and Other Approaches. Taylor & Francis, London, pp. 149—154.
Saari, J. and Lahtela, J., 1981. Work conditions and accidents in three industries. Scandinavian Journal of Work, Environment and Health 7: 97—105.
Saari, J., 1982. Summary of the results derived from the theme "Accidents and progress of technology". Journal of Occupational Accidents 4: 373—378.
Skiba, R., 1973. Gefahrenträgertheorie. Forschungsbericht Nr 106, Bundesanstalt für Arbeitsschutz und Unfallforschung, Dortmund, p. 47.
Smillie, R.J. and Ayoub, M.A., 1976. Accident causation theories: a simulation approach. Journal of Occupational Accidents 1: 47—68.
Smith, R.G. and Lovegrove, A., 1983. Danger compensation effects of stop signals at intersections. Accident Analysis and Prevention 15: 95—104.
Surry, J., 1968. Industrial accident research. University of Toronto, Toronto, p. 203.
WHO, 1982. Psychological factors in injury prevention. World Health Organization, OCH/83.4, Geneva, p. 31.

Wigglesworth, E.C., 1972. A teaching model of injury causation and a guide for selecting countermeasures. Occupational Psychology 46: 69—80.

Wilde, G.J.S., 1976. Social interaction patterns in driver behavior: an introductory review. Human Factors 18: 477—492.

Zimolong, B., 1976. Methoden der psychologischen und technischen Arbeitssicherheit. Psychologie und Praxis 20: 55—69.

Zimolong, B., 1979. Risikoeinschätzung und Unfallgefährdung beim Rangieren. Zeitschrift für Verkehrssicherheit 25: 109—114.

Zohar, D., 1978. Why do we bump into things while walking? Human Factors 20: 671—679.

APPLICATION OF HUMAN ERROR ANALYSIS TO OCCUPATIONAL ACCIDENT RESEARCH

W.T. SINGLETON

Applied Psychology Department, The University of Aston in Birmingham, College House, Gosta Green, Birmingham B4 7ET (United Kingdom)

ABSTRACT

Singleton, W.T., 1984. Application of human error analysis to occupational accident research. *Journal of Occupational Accidents*, 6: 107—115.

Currently there is increased interest in behavioural approaches to safety as distinct from the engineering approach which relies on physical separation of the person from the hazard. The objective of behavioural approaches is to reduce the probability of human error. This requires an understanding of the aetiology of human error and its role in accidents. The relative importance of environmental, informational, attitudinal and social factors needs to be assessed by the analysis of accidents in terms of the origins of the human errors.

The situation will be illustrated by the two extremes of errors; those which lead to damage only to the person making the mistake, as in agriculture, and those which in very rare instances might lead to damage to large numbers of people as in the process industries. The former can be a post-hoc accident analysis study but the latter is necessarily an anticipatory hazard and critical incident study.

INTRODUCTION

The Oxford Dictionary defines an error as a mistake following from a misunderstanding of a thing's meaning; it can occur in thought or in action. The Encyclopaedia of Psychology (Eysenck et al., 1975) defines an error as the difference between an observed and an expected value. At first sight these two concepts would appear to have little in common but it happens that they both fit well enough with the concept of error derived from skill theory.

A skilled person is conceived as interacting with the external world — including other people — by a continuous process of prediction and verification (Singleton, 1983a). Faced with a given situation, he will create an internal model of it and manipulate that model so as to arrive at an hypothesis as to what will happen if he takes a particular action. If this predicted outcome seems favourable in terms of his current intentions, he will then initiate the action and observe its consequences, in particular by comparisons with his prediction. If the new evidence from the senses confirms his prediction then all is well but if it does not then he has made an error. In the Dictionary terms he has misunderstood the situation, in the Encyclopaedia terms there is a discrepancy between the observed and the expected outcome.

Errors relate to safety because the unexpected outcome of human action may initiate a train of events which leads to damage to persons. On the other hand it must be recognised that errors are the fundamental source of information about the world. Hence the analogy of the human being with servo-mechanisms which are essentially error-actuated devices (Wiener, 1948). If a person takes an action which leads to events conforming precisely to his prediction then he will have made progress in relation to his intentions but technically he will have received no information and he will not be provoked into any learning. His skills will consolidate but not improve.

This is the origin of the dilemma of using people in high energy systems. The value of people as system components lies in their ability to learn but, for learning to take place, they must often make errors and these errors may result in energy leaks which in turn result in accidents. The obvious solution is to protect the operator (and the system) from his own actions by not allowing him to follow courses of action which might result in undesirable energy transmissions. This can be done by isolating him and other people as far as possible from the energy of the hardware system. This is the basis of the engineering approach to safety. Barriers are provided between the operator and the system energy — the simplest case is machine guarding. More sophisticated ways involve physically distancing the operator from the working components so that he is more remote from the action. Inevitably, there are snags. Economically it is often expensive and the safety specialist is faced with difficult cost/value decisions, where he must relate the cost of the isolating measures to the likely benefits of reducing the probability of accidents. Psychologically there is always a penalty in that the flow of sensory data to the operator will be reduced and distorted. There can be some compensation by the use of more elaborate instrumentation, e.g., by siting the operator in a control room filled with displays and controls of all kinds, but there is an inevitable reduction in contact with reality and there are now new sources of unreliability in the potential failure of information channels. This needs to be explored separately for simple and complex technical systems. As an example of a simple man—machine system consider the farm tractor; as an example of a complex man—machine system consider the nuclear power station.

THE FARM TRACTOR

One of the most common causes of serious accidents with tractors up to about ten years ago was overturning. Usually this happens when the driver attempts to turn on a slope in a field. It may be because he is going too quickly or the implement behind the tractor may run in the wrong direction in relation to the tractor itself ("jack knifing"). It also can happen on farm roads when the wheels on one side ride into a ditch or up a steep slope. The tractor can even overturn backwards when attempting to pull an immovable object such as a tree. Looked at from an engineering view point, the remedy is clear: the tractor must be fitted with a cab strong enough to resist crushing which will protect the driver should the tractor overturn. Most European countries now have legislation which makes such cabs compulsory. There is an incidental advantage in that the driver is protected from cold or wet weather. However there are also a number of disadvantages. If the weather is hot then the cab can be stifling; unless considerable care goes into acoustic features of design it can also be excessively noisy; and it interferes with ease of access to the driving seat. It also interferes with the flow of information to the driver, there are visual blind spots and he is less able to receive spoken messages or react to warning noises. In short he has been isolated from his physical and informational environments. The consequences of errors leading to one particular level of accident have been reduced but the potential for errors of other kinds which might lead to different kinds of accidents has been increased.

Taking the year 1981 as an example, there were 71 fatal accidents on U.K. farms; tractors were involved in 40 of these. Twelve of the 40 had apparently nothing to do with the cab — the most common incident was clothing caught in the P.T.O. shaft. Of the remaining 28, 12 were due to the overturning. Within this 12, 6 had no cab, 5 drivers were thrown out of the cab and 1 driver was killed by head injuries in the cab. That still leaves 16 fatal accidents possibly connected with cabs but not with overturning. The driver was crushed by the cab in 5 cases, he got into difficulties getting in and out of the cab in 9 cases and there were 2 fatal accidents which might have been avoided had the driver been in closer real contact with the surroundings (Fig. 1).

This can be compared with the situation ten years earlier. In 1971, there were 118 fatal accidents on farms of which 55 involved tractors. The cab or

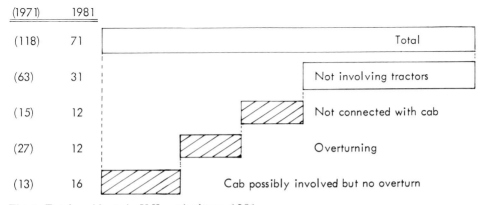

Fig. 1. Fatal accidents in U.K. agriculture, 1981.

lack of it was not involved in 15 of these. Of the remaining 40, 27 were overturnings and 13 were not. Of the 13, 4 involved crushing, 5 were to do with getting in or out and 4 connected with inadequate informational contact with the surroundings.

The interesting change over this ten year period is that there has been an important fall (from 27 to 12) in the number of fatalities due to overturning but in spite of a considerable fall in overall fatalities (118 to 71) there has actually been a rise (13 to 16) in the number of fatal accidents which might be traced to increased physical isolation — problems of getting in and out and the increased informational isolation and problems of not noticing something or someone. There have been safety benefits from the increased use of cabs but there have also been some penalties. The remedy would seem to lie in better ergonomic design of cabs for ease of access and egress and for ease of communications. Some studies of these issues are being made (e.g., Talamo and Stayner, 1983).

This is not a direct analysis of errors but rather an analysis directed by consideration of errors. The consideration of potential errors leads to knowing what to look for in the fatal accident data.

THE NUCLEAR POWER STATION

Here there are some error problems similar to those in agriculture in that maintenance engineers may make mistakes which damage themselves or the plant because of difficulties of access, inadequate visual information, high noise levels and so on (Seminara and Parsons, 1982).

However, the main focus of error studies is the control-room operator who is in a totally different situation. All the classical ergonomic variables have been taken care of. He works in an air-conditioned, well-lighted, low noise environment where every possible care has been taken of anthropometrics, dial design, control positioning, functional patterns of the interface, etc. He has high qualifications, extensive basic training and regular up-dating training by courses and simulators. His working hours and other conditions are carefully controlled. It is very unlikely that he will ever make an error which might result in immediate damage to himself. There is however a slight possibility — perhaps once every few years — that he will make a mistake leading to a plant shutdown, which will be expensive for his employer, and a very remote possibility — perhaps once in many thousands of years — that he will make a mistake which will lead to serious damage to the general public outside the plant (NUREG, 1975).

In these circumstances the study strategy is not to wait for accidents to happen and then analyse them in terms of errors so as to do better in the future, but rather to try to predict every conceivable possibility of error and find a way of guarding against it. This is a hazard analysis as distinct from an accident analysis. The aim is to maximise the system reliability including hardware, software and human system components. There is another fundamental difference from simpler systems in that the task of the operator is not so much to monitor the plant and stop it if things go wrong. The plant is very highly automated and it will invariably stop itself if anything serious goes wrong. Rather, the task of the operator is to engage occasionally in fine tuning and in providing alternative ways of functioning so that the plant is steered away from the possibility of an expensive unnecessary automatic shutdown. If the plant does shut down he has a further complex task in monitoring all the tidying up operations which again take place automatically to reduce the plant to a stable quiescent state. Incidentally, these trends are common to all process control situations and this is a new general role for the human operator in high technology (Singleton, 1983b).

There are many difficulties about system reliability data. Even for hardware components, failure rates are useful only in estimating such measures as the average maintenance load and the required size of spares stores; they give no indication as to when or why the plant will break down. Computer support systems are now so reliable and cheap (so that they can be multiplicated without excessive cost) that they give little operational trouble. Inevitably, the greatest uncertainty is about the human components. One procedure for coping with them is to try to estimate error rates for standard tasks (Swain and Guttman, 1983). This Technique for Human Error Rate Prediction (THERP) is based on analysing operations into elementary tasks which have standard error probabilities and then combing these probabilities with modifications by "performance shaping factors" such as stress so as to obtain estimates of error rates for particular events. This approach is nicely tuned to engineering thinking. It opens up the possibility that the human operator can be treated as just another system component within a series of fault-tree analyses of plant behaviour. Unfortunately it is impossible to operationally validate this technique (or indeed any other technique) because of the rarity of real traumatic events but, equally, the approach can not be relied upon as being psychologically sound. In short we have no means of knowing what the likely error rates in estimated error rates are. An alternative approach is to try to obtain and combine expert judgments about overall task reliability. This procedure is currently extremely popular. Akersten and Wirstad (1983) quote more than one hundred references, mainly over the period 1975—1983, about 20% of which are applications in nuclear power plants.

However, the accuracy of expert judgments is obviously a function of the expert's experience and again we come up against the barrier that there is very little experience of serious errors in the operation of nuclear power plants.

In attempting to systematise experience which becomes available steadily as plants continue to operate, it is desirable to have some means of categorising errors. A prerequisite for such a taxonomy is an understanding of how the operator functions in assessing situations and solving problems. There have been many attempts to describe these activities (Rasmussen and Rouse,

1981). There is an extensive speculation which could benefit from the discipline of more factual evidence but again methodology for empirically explaining these complex situations is currently very limited. Rasmussen (1982) distinguishes three levels of behaviour at each of which errors may occur: skill, rule- and knowledge-based performance (Fig. 2). This distinction recognises that there is a hierarchy of levels of behaviour but the terminology if unfortunate (skill can hardly be restricted to the lowest level of human performance) and the separation is rather too definitive. It might be more appropriate to label three particular levels as rules, principles and concepts (Fig. 3) and to recognise that these are not distinct categories but rather bench-marks along a continuous spectrum of styles of human activity. The

PERFORMANCE	MECHANISM	ERRORS
Skill-based	Stored patterns of behaviour (routines)	Variability of force, space or time coordination
Rule-based	Stored rules	Wrong classification or recognition. Erroneous associations
Knowledge-based	Data processing	Failures in reasoning about a casual network
	Selection of level of behaviour	Discrimination and choice

Fig. 2. Classification of performance levels and errors (Rasmussen, 1982).

LEARNING MODE	ERRORS
Following a drill	Unintended action sequence
Application of Rules	Failure of identification or memory
Use of Principles	Wrong classification
Conceptual Diagnosis	Creation of inadequate or wrong model

Fig. 3. Classification of skill levels and errors (Singleton, 1975).

kinds of errors shown in Fig. 2 are different but they do not necessarily relate to the particular mechanisms. The specified mechanisms are probably not mechanisms in the sense that they exist as different activities within the brain, rather they are labels for particular modes of learning appropriate for situations of different complexity. To operate by rules it is necessary to specify precisely the relevant informational stimulus situations where particular rules apply. Principles are generalisations which serve as guidelines — in other terminology, a map is available. Conceptual diagnosis involves generating a selective cognitive map before using it.

The test for these or other taxonomies is how well they fit in the classification of real incidents which have occurred in power stations. It may be disturbing to appreciate that, although there are about a hundred nuclear power stations operating across the world, the behavioural scientists are still only at the stage of devising methodology for error analysis. However it will be appreciated that most design or operational decisions can be made on established human factors principles. The design of control room physical and informational environments, the principles of training, etc. are quite well established (Pine et al., 1982; NKA/KRU, 1981). The increased understanding which should result from analysis of operational errors will contribute to detailed design decisions such as the relative importance of formal procedures versus strategic training, the kind and degree of operator decision support to be provided by computer driven displays, etc.

DISCUSSION

Error analysis is emerging as an oblique rather than a direct approach to the study of real situations. An understanding of how and why operators make mistakes is increasing rapidly at the present time — mainly stimulated by the operational needs of the process industries such as nuclear power. The obstacles which have inhibited progress are the misleading engineering procedures of treating the operator as just another system component and the absolutely fundamental nature of human error description within the behavioural sciences. Knowing about human error mechanisms is almost synonymous with knowing about human behaviour. The human operator is an integrated organism responding to a total situation. Data is supplied through the displays (real or instrumental) and he responds to the information which is generated by his interpretation of that data in the context of his knowledge of the total system, including his own sense of direction for the system. The errors which he makes generate further information which enables him to expand his model of the system and the way it functions.

It seems unlikely that further progress will be made by considering the operator as a complex but unchanging information processing device with specific capacities. For power stations this approach has served well in consideration of the "hard wire" man—machine interface. Given the increased flexibility available in computer-mediated displays and controls, together with the contextual complexity of the highly automated plant, it is useful to think of the operator as a continuously learning organism. Minor errors will need to be accepted within the system to facilitate further development of skill but major errors derived from inappropriate modelling must be avoided entirely. To this end the operator requires extensive support from regulations, procedures, computer-based simulations and colleagues. The design of these support systems cannot be completed ab initio. The interface must be modified continuously in the light of experience. Experience is essentially experience of errors. This needs to be systematised by routine collection of operational data about errors classified in various ways. The error classifications themselves need continuous development. An error should not be conceptualised as something which resulted in an accident nor as something which indicates an operator's inadequacy but, rather, as a system-based deficiency which might be avoided in future by a design or training modification.

It is not possible to go into such details for errors in tractor operation. In this case accidents can be analysed in terms of why the person or people involved made particular kinds of errors which define particular kinds of accidents. Note that this also is a change of orientation in that accidents are not to be categorised in terms of the physical incident but rather in terms of particular behavioural patterns which might happen to result in a variety of physical incidents. For example, head injuries might be a homogeneous category from a medical viewpoint but behaviourally a head injury due to falling off a ladder is a different accident problem from a head injury due to a falling object.

Tractors, power plants and many other machines exhibit in common a particular design trend which creates new behavioural problems. In the wake of technological developments, the operator is increasingly isolated from the physical world with which he must interact and the information he gets is more and more channelled into the visual system. It is perhaps this trend which lies behind the disturbing phenomenon which has been variously labelled as "cognitive narrowing", "fascination", "perceptual isolation", etc. Typically, the operator behaves as though he has an intellectual problem which has been presented to him visually and he appears to forget that that problem is to do with real goings on out there in the physical world. Moreover he develops an hypothesis as to what lies behind the particular pattern of visual signals he is receiving and he sticks obsessively to this hypothesis regardless of more and more data presentations which ought to cause him to radically revise his hypothesis. This is obviously very dangerous. The remedies include trying to provide relevant data for other sensory channels — hearing, touch and kinesthesis — and ensuring that the operator does have regular direct contact with the real physical system which he controls remotely.

The approach to errors taken in this paper is an idiosyncratic one in that the underlying theory utilised is essentially developmental, phrased in terms of human skill. It is possible and reasonable to consider errors as arising from equivocation and ambiguity in the human information channel or as failures in the activity of responsible individuals or as indicators of inadequate selection criteria, etc. Equally, there are many other ways of conceptualising accidents. It seems unlikely that any one approach will suffice to cope with the variety of human behaviour in work situations but the "errors within skilled performance" concept outlined in this paper should merit a place within the total repertoire of behavioural approaches to safety.

REFERENCES

Akersten, P.A. and Wirstad, J., 1983. Expert judgement for safety work in nuclear power plants. Ergononamrad A.B. Report No. 23.
Eysenk, H.J., Arnold, W.J. and Meili, R. (Eds.), 1975. Encyclopedia of Psychology. Fontana, London.
NUREG, 1975. Reactor safety study. An assessment of accident risks in U.S. commercial nuclear power plants. U.S. Nuclear Regulatory Commission. NUREG 75/014.
NKA (KRU), 1981. Technical summary report on control room design and human reliability. Nordic Liaison Committee for Atomic Energy. NKAKRU (81) 63.
Pine, S.W., Schulz, K.A., Ledman, T.R., Hanson, T.G. and Evans, T.G., 1982. Human Engineering Guide for Enhancing Nuclear Control Rooms. Electric Power Research Institute, Palo Alto. EPRI, N.P. 2411.
Rasmussen, J. and Rouse, W.B., 1981. Human Detection and Diagnosis of System Failures. Plenum, New York.
Rasmussen, J., 1982. Human errors. A taxonomy for describing human malfunction in industrial installations. Journal of Occupational Accidents. 4: 311—334.
Seminara, J.L. and Parsons, S.O., 1982. Nuclear power plant maintainability. Applied Ergonomics, 13 (3): 177—189.
Singleton, W.T., 1975. Skill and accidents. In: Occupational Accident Research, Stockholm, Swedish Work Environment Fund.
Singleton, W.T. (Ed.), 1983a. The Study of Real Skills. Vol. 4. Social Skills. M.T.P. Press, Lancaster.
Singleton, W.T., 1983b. The skilled operator as the arbiter in potential accident situations. Paper to the Xth World Congress on the Prevention of Occupational Accidents and Diseases, Ottawa.
Swain, A.D. and Guttman, H.E., 1983. Handbook of Human Reliability Analysis with Emphasis on Nuclear Power Plant Applications. (Final Report) NUREG/CR — 1278.
Talamo, J.D.C. and Stayner, R.M., 1983. The requirement of tractor drivers to hear sounds produced external to the tractor cab. National Institute of Agricultural Engineering. Engineering Design & Development Division. D.N. 1189.
Wiener, N., 1948. Cybernetics. Wiley, New York.

THE ROLE OF DEVIATIONS IN ACCIDENT CAUSATION AND CONTROL

URBAN KJELLÉN

Occupational Accident Research Unit, Royal Institute of Technology, S-100 44 Stockholm (Sweden)

ABSTRACT

Kjellén, U., 1984. The role of deviations in accident causation and control. *Journal of Occupational Accidents*, 6: 117—126.

A thorough understanding of the role of deviations in accident causation improves the possibilities of a systematic feedback control of accidents in production systems in operation. A systems variable is classified as a deviation when the variable takes values that fall outside a norm.

A statistical model of relations between deviations and accidents is presented. Results of a review of empirical research on these relations and on the collection and use of data on deviations in accident control are discussed.

It is concluded that certain types of deviations are valid indicators of the risk of accidents in the context of the production systems in which they occur. The practices inside companies for accident and near-accident reporting and for safety inspections do not provide reliable and comprehensive data on deviations needed for accident control. There is thus a need for development and improvement of these practices in order to utilize the full potential of the information on hazardous deviations that is available at workplaces.

INTRODUCTION

In Sweden, about 200,000 safety representatives, supervisors, and safety engineers are engaged in safety activities. These activities include investigations and follow-ups of accidents and near-accidents, safety inspections, and reviews of new machinery and production systems. There is in general a lack of positive scientific evidence of the effects of these activities. For example, more than 100,000 Swedish employees are injured at work each year due to an occupational accident. This figure has remained more or less unchanged during the last decade (Statistics Sweden, 1983).

In an early study of the Occupational Accident Research Unit (OARU), the information systems for feedback control of accidents of six Swedish companies — i.e. accident and near-accident reporting, and safety inspections — were evaluated (Kjellén, 1982). The study revealed a number of shortcomings of these different safety information systems.

For example, the data documented in lost-time and near-accident reports and in safety inspection protocols were filtered; i.e. certain types of information on accident risks were suppressed. The various filters narrowed the scope for the type and timing of corrective and preventive measures. Further, there was a general lack of routines for the distribution of information on accident risks and preventive measures to the various functions of the company outside the safety organization; i.e. higher management, staff officers, and workers.

NEED FOR AN ANALYTICAL FRAMEWORK OF ACCIDENTS

There are several explanations to account for these deficiencies. One plausible explanation is the lack of adequate tools in the form of analytical frameworks of accidents for use inside companies in safety practice and in the training of safety practicians. A tool of this type has two basic potentials:

(1) It has potential for cognitive change; i.e. shaping the perception of the nature of accidents of those involved in or affected by safety activities. An accident perception that has been incorporated into the intuitive modes of thinking of a safety practitioner will affect his behaviour in relation to accidents in general; it will in particular affect his inquiries into accident causal factors in specific situations, his conclusions from these inquiries, and his choice of remedial action.

(2) It has mechanical potential; i.e. to serve as a basis for the design of questionnaires, forms, checklists, and procedures.

It follows, that the application of an analytical framework of accidents in safety practice affects the possibilities of, among other things, a comprehensive and reliable charting of accident causal factors.

THE OARU MODEL

Several different theories and models of accidents are presented in the research literature. Selective theories and models have been translated into analytical tools or frameworks for application in investigations of accidents. A scrutiny by the OARU of different existing analytical frameworks resulted in the conclusion that these were too complex, too abstract or too narrow in scope for application by safety practicians inside companies.

A model of accidents was designed which represents a further development and synthesis of a number of existing and proven models (Kjellén and Larsson, 1981). The model was further developed and translated into an analytical framework and a procedure for investigation into accidents. The analytical framework and the procedure have been tested by safety practicianers of, for example, building companies and paper mills (Harms-Ringdahl, 1983; Kjellén, 1983a).

According to the model, an accident process is made up of a flow or pattern of causally and chronologically related deviations. A *deviation* is the classification of a systems variable, when the variable takes values that fall outside a norm. The occurrence of the first deviation in the pattern of deviations denotes the initiation of the accident process, which terminates when the body has fully absorbed the energy that caused damage to body tissue.

The accident process is further characterized by the transition of a production system from a state of full or partial control of the energies of the system to a state of loss-of-control of these energies and the subsequent personal injury. Deviations which occur prior to the loss-of-control of the energies of the system are termed initial-phase deviations. A concluding-phase deviation is an event of the uncontrolled energy flow of the system that follows the loss-of-control of the system.

The OARU model also includes the concept of *determining factors*, i.e. technical, human or organizational resources of the production system that affect the occurrence or consequences of deviations.

MOTIVES FOR STUDYING DEVIATIONS

There are several reasons for introducing the term "deviation" as a central concept of the OARU model (Kjellén, 1983b). A review of different terms of the occupational accident research literature has shown that the application of most of these terms presupposes the existence of norms (Kjellén, 1983c). Examples of such terms are "near accident", "critical incident", "unsafe act", "unsafe condition" and "disturbance". Various types of norms appear in the literature, for example:

- standard, rule or regulation;
- adequate or acceptable;
- normal or usual;
- expected, planned or intended; and
- homeostasis.

A characteristic of each of these different types of norms is the process through which the norm is established. The norm serves as a decision criterion or filter in the collection of data from the industrial production process about accident risks. The filter is essential in order to reduce the potential volume of information about accident risks to a manageable size. The above mentioned types of norms suffer from problems of limited applicability and/or problems of fuzziness when applied in data collection. For example, opinions may differ between workers on the one hand and management and systems designers on the other as to what constitutes a faultless production process. This difference in opinion may in itself constitute an accident risk. One motive for introducing the deviation concept was to develop more precise operational definitions of terms used in safety practice. This is accomplished by stimulating the group processes inside companies that result in the identification and resolution of such differences through the establishment of norms that produce reasonable risk levels. Consequently, the term deviation is defined in relation to the norms for the planned production process.

A second motive for introducing the term "deviation" was to let this term act as a tool in change processes, including the evolution of more valid and comprehensive accident perception, in members of industrial organizations.

Third, the deviation concept is related to a strategy for accident prevention. This strategy emphasizes the local characteristics of the phenomenon of accidents and the needs for adaptive response at the company and department levels to changing conditions. One purpose was to widen the scope for accident prevention to include all management systems for production control. Consequently, a taxonomy of deviations has been developed, which is related to various traditional systems of production planning and control (Table 1). A further purpose was to emphasize the importance of long-term preventive measures as an indispensible complement to the short-term corrective actions (i.e. the elimination of deviations). Long-term measures relate to the design and development of the production system. These include measures which aim at reducing or eliminating the probability of the recurrence of deviations and/or the expected harmful consequences of deviations.

TABLE 1

Examples of relations between classes of deviations and different traditional systems of production planning and control

Class of deviations	System of production planning and control
A. Flow of material	Material control
B. Flow of labour force	Personnel control
C. Flow of information	Supervision
D. Technical	Technical control
E. Human	First-line supervision
F. Intersecting/parallel activities	Activity control
G. Environmental factors	Health and safety activities
H. Stationary guards	Health and safety activities
I. Personal protective equipment	Health and safety activities

Fourth, the deviation concept is related to a method or procedure for the collection, processing and distribution of information inside companies for accident control purposes. The procedure is rooted in cybernetics and in the process of diagnosis and includes the following steps: (1) collection of data on deviations in accident and near-accident investigations and in safety inspections; (2) analysis of the data in investigation groups with respect to systems failures (i.e. determining factors) and safety measures; (3) decisions to implement measures; and (4) follow-up of the effects of implemented measures.

Fifth, the deviation concept is applicable to a wide range of production systems and accident types. In particular, it provides a heuristical method for the solving of complex, ill-structured accident problems.

The application of the deviation concept in local safety activities is based on the assumption that the level of safety of a production system, and the extent to which the system is controllable at the department and company levels, are related. This assumption is supported by evidence from earlier research. One study showed a positive relation between safety and the degree of management control of production costs (Grimaldi, 1970). Accident research has also shown that the risk of accidents is higher in non-productive tasks, which are usually less well planned in advance (Saari and Lahtela, 1981).

A STATISTICAL MODEL OF THE RELATION BETWEEN DEVIATIONS AND THE RISK OF ACCIDENTS

The general definition of a deviation according to the OARU model applies to a great variety of features of production systems. In safety practice, it is necessary to focus on a limited number of critical deviations; i.e. deviations which significantly affect the risk of accidents and, in addition, are measurable and controllable.

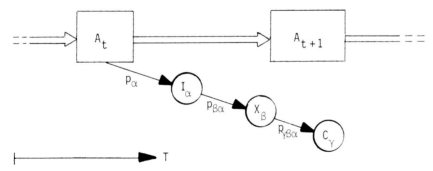

Fig. 1. Risk assessment model.

A statistical model for risk assessment is shown in Fig. 1 (see Kjellén, 1983c, for a detailed description). The aim of the model is to identify the components that are necessary to assess the significance of deviations with respect to the risk of accidents. The model examines the stochastic process by which a planned activity of a production system, A_t, develops into an uncontrolled sequence of events with subsequent harm. An incident, I_α, is made up of an uncontrolled energy flow; i.e. a single concluding-phase deviation or a sequence of causally related deviations of this type. The outcome of the incident is denoted X_β. This variable takes two values; i.e. X_0

(energy flow is short of body injury thresholds) and X_1 (energy flow exceeds body injury thresholds). A given consequence of an injurious outcome, X_1, is denoted C_γ, where γ is an index indicating a degree of severity (e.g., number of workdays lost).

The risks of accidents, $\hat{p}_t$, for an activity A_t of the production system equals:

$$\hat{p}_t = \sum_{\gamma=1}^{6000} \hat{p}_\gamma = \sum_{\gamma=1}^{6000} \sum_\alpha p_\alpha \cdot p_{1\alpha} \cdot p_{\gamma 1\alpha} \qquad (1)$$

p_α is the probability of the occurrence of the incident I_α. $p_{\beta\alpha}$ is the conditional probability of the outcome X_β of this incident. $p_{\gamma\beta\alpha}$ is the conditional probability of the consequence C_γ of the incident I_α and the outcome X_β. $\gamma = 6000$ corresponds to a permanent disability or fatality (ANSI, 1967). The characteristics of I_α and the conditional probabilities p_α, $p_{\beta\alpha}$, and $p_{\gamma\beta\alpha}$ are parameterized by the parameters describing A_t.

The significance of incidents and concluding-phase deviations

An incident is, according to the definition applied in this model, a necessary condition for the occurrence of an accident. An example of a measure of the *significance of an incident*, I_α, as an indicator of the risk of accidents is the probability that this incident results in a lost-time accident, i.e.:

$$\hat{p}_{t;\alpha} = \sum_{\gamma=1}^{6000} p_{1\alpha} \cdot p_{\gamma 1\alpha} \qquad (2)$$

According to an early study, the average value of $\hat{p}_{t;\alpha}$ for different types of individuals, incidents and production systems equals 0.003 (Heinrich, 1959).

Researchers have argued that the consequences of incidents are largely fortuitous (see, e.g., Tarrants, 1963). This should imply that $\hat{p}_{t;\alpha}$ is independent of the type of incident and context. However, empirical studies contradict this hypothesis. It has been shown that the significance of an incident is dependent on the type of incident (i.e. pattern of concluding-phase deviations) and context in which it occurs, for example, with respect to the energy involved (Shannon and Manning, 1980; Heinrich et al., 1980; Schneider and Krause, 1969).

The significance of initial-phase deviations

The probabilities p_α, $p_{\beta\alpha}$, and $p_{\gamma\beta\alpha}$ are functions of systems variables V_l. The set or range of values of a systems variable is dichotomized into two classes; normal (N_l) and deviation (D_l). The existence of an operationally defined norm for each variable V_l, that separates these two classes is assumed.

$\hat{p}_{t;D_l}$ denotes the probability of a lost time accident if $V_l = D_l$. An initial-phase deviation (D_l) is hazardous, i.e. is a valid indicator of the risk of accidents, if $\hat{p}_{t;D_l} > \hat{p}_{t;N_l}$. There are many different combinations of the relations $(p_{\alpha;D_l} > p_{\alpha;N_l})$, $(p_{1\alpha;D_l} > p_{1\alpha;N_l})$, and $(p_{01\alpha;D_l} < p_{01\alpha;N_l})$ that will make this relation valid.

For example, a deviation such as the absence of a required guard on a power saw (D_g) will increase the probability of an accident, given the incident, I_a, i.e. the guide bar has come loose and the operator's hand slipped and moves in direction of the rotating saw blade. That is $(p_{1a;D_g} > p_{1a;N_g})$. This deviation will thus increase $\hat{p}_t$ if no other probabilities are changed.

In production systems, there is often a probability associated with a systems variable V_l taking the value D_l during an activity A_t. This probability is denoted $\bar{\bar{p}}_l$. The expected value of $\hat{p}_t$ as a function of the probability $\bar{\bar{p}}_l$ equals:

$$E\hat{p}_t(\bar{\bar{p}}_l) = \hat{p}_{t;N_l} \cdot (1 - \bar{\bar{p}}_l) + \hat{p}_{t;D_l} \cdot \bar{\bar{p}}_l = \hat{p}_{t;N_l} + \bar{\bar{p}}_l \cdot (\hat{p}_{t;D_l} - \hat{p}_{t;N_l}) \qquad (3)$$

If $\hat{p}_{t;N_l} < \hat{p}_{t;D_l}$ a reduction of the probability $\bar{\bar{p}}_l$ of a deviation D_l will reduce the expected probability of a lost-time accident $\hat{p}_t$. The expression $(\hat{p}_{t;D_l} - \hat{p}_{t;N_l})$ is a measure of the significance of a deviation with respect to the risk of accidents.

Empirical studies of this relation are based on analytical or statistical methods. *Analytical methods* (e.g. fault tree analysis) makes it possible to establish logical relations between initial-phase deviations (D_l) on the one hand and incidents (I_α), outcomes (X_β) and consequences (C_γ) on the other. For example, there are four different types of logical relations between initial-phase deviations and incidents (cf. Monteau, 1977):

(1) D_l is a necessary and sufficient condition for the occurrence of I_α, i.e. $p_{\alpha;D_l} = 1$ and $p_{\alpha;N_l} = 0$.

(2) D_l increases the probability of I_α, i.e. $p_{\alpha;D_l} > p_{\alpha;N_l}$.

(3) The occurrence of I_α is independent of the occurrence of D_l, i.e. $p_{\alpha;D_l} = p_{\alpha;N_l}$.

(4) D_l is a necessary but not sufficient condition for the occurrence of I_α, i.e. $p_{\alpha;D_l} > 0$ and $p_{\alpha;N_l} = 0$.

By applying logical analysis to studies of accidents of certain production systems, deviations have been identified which fulfilled condition (1), (2) or (4); i.e. were valid indicators of the risk of accidents of the production systems concerned (see Monteau, 1977). These included, for example, replacement of workers, unsuitable method of operation, absence of personal protective equipment, and faulty equipment.

Case-comparison and correlation studies are examples of *statistical methods* which are applicable to studies of relations between deviations and accidents. In case-comparison studies, accident-stricken activities are compared with a random sample of accident-free activities with respect to certain variables. Examples of variables that, according to two studies, were significantly more common in accident-stricken activities are defective equipment, high/low/irregular work load, untrained workers (Hagbergh, 1960), and production disturbances and non-ordinary workplace (Saari, 1976/77). The values of variables that were identified in these two studies as being related to the risk of accidents are frequently regarded as unplanned (i.e. deviations) by members of industrial organizations.

Safety or activity sampling is a method of measuring the frequency of initial-phase deviations with a relatively high degree of reliability (see, e.g., Komaki et al., 1978). A study based on this method has shown that the patterns of injury and incident frequency and the patterns of deviation frequency were correlated (Meyer, 1963). Other studies attribute a reduction of the frequency of certain types of deviations and a simultaneous reduction of the accident frequency rate to the application of safety sampling in the control of these deviations (Komaki et al., 1978; Rees, 1967).

The above mentioned studies do not generate scientific evidence of general relations between initial-phase deviations and accidents. This is, for example, due to: (1) problems of interpreting research results because of a non-randomized study design; (2) the subjective element of the definition of deviations that makes comparisons between separate studies difficult; and (3) the contextual dependence of relations between deviations and accidents.

It is uncertain whether it ever will be possible to identify general relations between type of deviation and the risk of accidents. However, this fact does not disqualify the deviation concept as a suitable basis for the development of practical tools to be used inside companies in the systematic prevention of accidents.

COLLECTION OF DATA ON DEVIATIONS

Data are collected and analyzed in safety practice for the purpose of identifying and assessing accident risks and of developing and evaluating safety measures. As has been mentioned above, data from workplaces about accident risks must be filtered and organized in order to be useful as a basis for accident control.

In practice, such filtering and organization actually takes place (see the Introduction). They are affected by a number of conditions which are only to a minor extent subject to conscious control by the industrial organization. The practical significance of the deviation concept is its potential to bring about a consciously controlled filtering and organization of data from the production system.

On the basis of the risk assessment model of the previous section, three different types of needs for data on deviations can be identified: (1) descriptive data on initial-phase deviations (D_l) and incidents (I_α); (2) the probabilities of incidents (p_α), outcomes ($p_{\beta\alpha}$) and consequences ($p_{\gamma\beta\alpha}$), given the presence and absence respectively of initial-phase deviations (D_l); and (3) the probability of initial phase deviations ($\bar{\bar{p}}_l$).

Data on deviations are collected in accident and near-accident reporting and in safety inspections. There are two main problems associated with these three types of methods for data collection; i.e. the systematic suppression or filtering of data on certain types of deviations and incidents and unreliability in the documentation of deviations and incidents. The various filters prevent the comprehensive mapping and description of deviations and incidents. The routine reporting of accidents and near-accidents and routine safety inspection inside companies usually do not generate reliable data for the estimation of the probabilities of initial phase deviations, incidents, outcomes and consequences. This holds true, for example, for the average probabilities of incidents, outcomes and consequences, as well as for the marginal distributions of these probabilities, given the presence and absence of initial phase deviations.

In various research studies, it has been possible to improve the comprehensiveness and reliability in data collection. For a review of this research, see Kjellén (1983c). The measures that have been introduced include the application of checklists, special training of data collectors, feedback of results to the data collector and fixed routines. In addition, improvements in near-accident reporting have been accomplished through anonymous reporting, reporting limited to certain well-defined types of near accidents, and reporting during a limited time period.

Further research will show whether it is possible to develop reliable and valid predictors of the accident risk of specific industrial systems on the basis of the occurrence of deviations in these systems. This would require improvements of the current corporate safety information systems and a long-term systematic accumulation of experience, for example, through the aid of computers.

CONCLUSIONS

The theoretical analysis of the deviation concept as well as results of empirical research give support to the conclusion, that this concept provides a valid basis for the design of corporate information systems for the efficient control of accidents. The deviation concept is applicable to a wide range of production systems and accident types. Further, the concept provides an opportunity for the integration of production and safety management.

In practice, it must be acknowledged that there is a general lack of scientific data on the validity or significance of deviations with respect to the risk of accidents. Inquiries about the significance of deviations with respect to the risk of accidents are made through a combination of logical analyses and judgements made by members of the industrial organization.

REFERENCES

ANSI, 1967. Method of recording and measuring work injury experience: Z-16.1. American National Standards Institute, New York.
Grimaldi, J.V., 1970. The measurement of safety engineering performance. Journal of Safety Research 2: 137—159.
Hagbergh, A., 1960. Olycksfall, individ och arbetsmiljö. Personaladministrativa rådet, Report No. 23, Stockholm.
Harms-Ringdahl, L., 1983. Learning by accident. (In Swedish). Royal Institute of Technology. Report No. Trita-A0G-0025, Stockholm.
Heinrich, H.W., 1959. Industrial Accident Prevention. McGraw-Hill, New York.
Heinrich, H.W., Petersen, D. and Roos, N., 1980. Industrial Accident Prevention. McGraw-Hill, New York.
Kjellén, U., 1982. An evaluation of safety information systems of six medium-sized and large firms. Journal of Occupational Accidents 3: 273—288.
Kjellén, U., 1983a. The application of an accident process model to the development and testing of changes in the safety information systems of two construction firms. Journal of Occupational Accidents 5: 99—119.
Kjellén, U., 1983b. Analysis and development of corporate practices for accident control. Unpublished doctoral dissertation. Royal Institute of Technology, Report No. Trita-AVE-0001, Stockholm.
Kjellén, U., 1983c. The deviation concept in occupational accident control — theory and method. Royal Institute of Technology. Report No. Trita-A0G-0019, Stockholm.
Kjellén, U. and Larsson, T., 1981. Investigating accidents and reducing risks — a dynamic approach. Journal of Occupational Accidents 3: 129—140.
Komaki, J., Barwick, K.D. and Scott, L.R., 1978. A behavioral approach to occupational safety: pinpointing and reinforcing safe performance in a food manufacturing plant. Journal of Applied Psychology 63(4): 434—445.
Meyer, J., 1963. Statistical sampling and control for safety. Industrial Quality Control, June.
Monteau, M., 1977. A practical method of investigating accident factors. Principles and experimental application. Commission of the European Communities, Luxemburg.
Rees, A.G., 1967. Safety sampling — a technique for measuring accident potential. The British Journal of Occupational Safety 7(79): 190—195.
Saari, J., 1976/77. Characteristics of tasks associated with the occurrence of accidents. Journal of Occupational Accidents 1: 273—279.
Saari, J. and Lahtela, J., 1981. Work conditions and accidents in three industries. Scandinavian Journal of Work Environment Health 7: suppl. 4, 97—105.
Schneider, B. and Krause, H., 1969. Zum Informationswert nicht meldepflichtige Unfälle für die betriebliche Unfallverhütung. Die Berufsgenossenschaft Betriebssicherheit, January: 7—10.
Shannon, H. and Manning, D., 1980. Differences between lost-time and non-lost-time industrial accidents. Journal of Occupational Accidents 2: 265—272.
Statistics Sweden, 1983. Occupational Injuries 1980. Liber Förlag, Stockholm.
Tarrants, W., 1963. An evaluation of the critical incident technique as a method for identifying accident causal factors. Unpublished doctoral dissertation, New York University, New York.

ABSTRACTS

Accident Models: How Underlying Differences Affect Workplace Safety

LUDWIG BENNER, JR.

University of Southern California, 12101 Toreador Lane, Oakton, CA 22124 (U.S.A.)

This paper reports results of an inquiry into accident models and investigation methodologies driving accident investigation programs in 17 U.S. governmental organizations to determine how they influenced the safety problem discovery and definition processes in those organizations. 14 accident models and 18 investigation methodologies were identified. The uses of these models and methodologies were then examined. From their influence on the functioning and results of the organizations' safety problem discovery and definition processes, criteria for judging the merits of each accident model were isolated and defined. These criteria included:

1. REALISTIC accident representations
2. DEFINITIVE AI data requirements
3. DIRECT problem definition
4. COMPREHENSIVE scoping of accident
5. DISCIPLINING for investigators's tasks
6. CONSISTENT with safety concepts
7. SATISFYING statutory missions
8. FUNCTIONAL links to task design
9. NON-CAUSAL to avoid abuses
10. VISIBLE and sensible to laymen

The criteria were then used to rate the observed models.

Rank	Rating
1. Events process model	19
2. Energy flow process model	18
3. Fault tree model	14
4. Haddon matrix model	8
5. All-cause model	7
6. Mathematical models	7
7. Abnormality levels model	7
8. Personal models	5
9. Epidemiologic model	4
10. Pentagon explosion model	4
11. Stochastic variable model	3
12. Violations model	3
13. Single event + cause factors model	1
14. Chain-of-events (c-o-e) model	1

The impacts of the investigative methodologies resulting from these models were examined next, and criteria for judging the methodologies were isolated and defined.

The criteria for the methodologies included:

1. HARMONIOUS participation
2. BLAMELESS outputs
3. PERSUASIVE descriptions
4. TIMELY problem discovery
5. EFFECTIVE results
6. DEFINITIVE corrections
7. valid EXPECTANT norms
8. SELF-TESTS for truth
9. REPLICABLE work products
10. LOOP-CLOSING validations

These criteria were then used to rate the methodologies.

Rank	Rating
1. Events analysis	18
2. MORT system	18
3. Fault tree	16
4. NTSB board + interorganizational groups	12
5. Gannt charting	12
6. Interorganizational multidisciplinary group	11
7. Board, with intraorganizational groups	10
8. Personal/good judgment	10
9. Baker police system	9
10. Epidemiological	9
11. Kipling's 5 w's + how	8
12. Closed-end flow charts	6
13. Compliance inspection	6
14. Statistical data gathering	5
15. Find chain-of-events (c-o-e)	4
16. Fact-finding/legal (JAG)	4
17. Complete the forms	3
18. Find cause, fix it	3

Accidents were then "reinvestigated" with the highest-rated models and methodologies to determine what improvements they might offer over existing practices. The results of retrospective applications of the preferred models to reports prepared under various models indicate that the differences in results are as substantial as the ratings would suggest. In the worst case, not a single valid descriptive fact about the accident phenomenon was included in a report driven by the violations model. Other findings included incomplete scoping; misdirected or unjustified conclusions and allegations; technical errors and omissions; delayed problem discovery; undetected problems; compromised citations; reinvestigations; and overlooked management and supervisory roles. The evidence is compelling that preferred accident models and investigation methodologies should be given priority attention to improve safety problem discovery and correction processes. Their use has already suggested basic analytical units for building compatible technical models to serve predictive safety analysis, accident investigation, task design, equipment design, task monitoring and other functional demands, and has suggested new accident information system approaches. Other improvements are predicted.

Serious Occupational Injuries with Special Regard to the Lack of Risk Control

JAN THORSON

The Swedish Foundation for Occupational Safety and Health of State Employees, Västra Esplanaden 19B, S-902 48 Umeå (Sweden)

Regarding subjective causes of injuries it was documented that negligence is more important than misadventure. This condition should be regarded with considerable interest from the point of view of injury-minimizing measures. First, responsibility for the injurious environments should be analysed in relation to existing standards. The result should be used for long-term prevention. Changes to the criminal law were made on the basis of experiences of this kind in Sweden in 1980.

As to objective causes, the main problem is lack of passive measures to control injury risks by impacts in road and other traffic.

Use of Case Reports from Work Accidents

JENS RASMUSSEN

Risø National Laboratory, DK-4000 Roskilde (Denmark)

In contrast to traditional statistical analysis of work accident data, which typically gives very general recommendations, the method proposed for analysis of case reports identifies very explicit countermeasures. The method focuses upon identification of sensitivity to future improvements rather than explanation of past accidents. Improvements require a change in human decisions during equipment design, work planning, or the execution itself. The use of a model of human cognitive functions is proposed for identification of the human decisions which are most sensitive to improvements.

Epidemiology of Occupational Accidents

BÖRJE BENGTSSON

National Board of Occupational Safety and Health, S-171 84 Solna (Sweden)

Epidemiology is the science of diseases in the population. Thus, this definition does not involve accidents. However, an approach similar to the one which is used in epidemiology may also be applied in the field of occupational accidents.

In the present paper the possibilities and limitations of the epidemiological approach as regards accidents will be discussed. Some examples of situations in which the approach may be useful in occupational accident research are given. One example concerns an attempt to calculate relative risks of accidents for different types of machine tools.

On Hand Injuries

TORE J. LARSSON

Occupational Accident Research Unit, Royal Institute of Technology, S-100 44, Stockholm (Sweden)

The venture of connecting injury-types with typical accident processes is heavily dependent upon a sound method of describing accident sequences.

In the project Sweden's two most comprehensive data sources on occupational accidents are pooled and cross-computed and 1612 permanently disabling hand injuries are used to probe the above-mentioned issue of prediction and investigate whether it is possible to locate the risk-groups and also envisage typical industrial accident-processes leading to severe hand injuries.

A Tentative Conceptual Analysis of Safety Activity

NED CARTER

National Board of Occupational Safety and Health, S-171 84 Solna (Sweden)

Much human behavior occurs due to the effects of consequences of similar behavior in previous situations. It is hypothesized that, in general, safety activities are performed in order to avoid the negative consequences which would otherwise result. However, infrequently occurring negative consequences will probably not maintain safety behavior nor will they readily compete with the effects which positive consequences have on other behavior. The consequences of this analysis for the design, implementation and maintenance of safety activities are discussed.

Validity and Utility of Experimental Research on Severe Work Vehicle Accidents

LENNART STRANDBERG

Accident Research Section, National Board of Occupational Safety and Health, S-171 84 Solna (Sweden)

Vehicles are involved in almost every other fatal occupational accident, even if commuting accidents are disregarded. Most of the fatalities occur with moving vehicles. Vehicle safety improvements have therefore been concentrated on the prevention either of dangerous vehicle motions (active safety) or of injuries upon impact (passive safety).

Since retrospective accident analyses cannot provide sufficient details for vehicle and environmental improvements, a number of methods have been developed for research on critical phenomena — by experiments in the field and in the laboratory, as well as by computer simulation.

Various applications require different degrees of complexity in the experimental models of human behaviour, of vehicle subsystems and of the environment, if valid data is to be procured without the experiments inflicting unacceptable damage or injuries. Unfortunately, ignorance of driver—vehicle dynamics as well as superficial "common-sense" models have led to large-scale production of hazardous designs, such as articulated vehicles, rear wheel driving and steering, etc.

The utility of some experimental techniques is illustrated with applications using different types of vehicles. Comparisons between data from model experiments and from measurements on heavy vehicles are used as an example of the validation procedure, which is essential for the credibility of models.

Worker Safety among Wage-Earners — The Connection between Individual Factors, Working Conditions and Industrial Accidents

EILA RIIKONEN

Tampere Regional Institute of Occupational Health, Box 486, SF-33101 Tampere 10 (Finland)

The aim of the present study is to locate the factors at the individual and workplace level that underlie industrial accidents and discuss their interrelations. A system model involving a strain model is used as the frame of reference.

The material (collected by the Central Statistical Office) consists of interviews of 5778 wage-earners, of whom 510 had had industrial accidents during the 12-month period before the interview.

Workers in industry, agriculture and forestry had most accidents. While both timeworkers and pieceworkers fell into the most accident-prone category, noise, exhaust gases, heavy lifting and poor posture tended to increase the accident risk.

Hazards in Stationary Grinding Machines

RISTO TUOMINEN and MARKKU MATTILA

Tampere University of Technology, P.O. Box 527, SF 33101 Tampere 10 (Finland)

Hazards in stationary production grinding machines in the Finnish metal industry have been studied systematically in literature reviews, accident analysis and supplementary hazard analysis. Accident, hygiene, and loading hazards were identified.

The analysis of 69 accident reports showed that in most accidents the injuring energy is connected with the necessary, controlled operation of the machine. The injuring energy can be regarded as an ordinary and planned characteristic of the technical work process. The operator must avoid this energy as a part of his work requirements.

The majority of accidents involved secondary tasks of the operator. It is typical that these tasks are performed in the hazardous area. Very high requirements and limits are placed on the proper behaviour of man in the process. Contributing factors, such as noise, dust and poor work practices, have a great influence on accident rates as well.

Fulltime isolation of the working zone seems to be needed as a priority preventive measure of safety engineering.

RESEARCH APPLICATION OF OCCUPATIONAL INJURY DATA SYSTEMS AT NATIONAL OR BRANCH LEVEL

DESCRIPTIVE EPIDEMIOLOGY IN JOB INJURY SURVEILLANCE

PATRICK J. COLEMAN

National Institute for Occupational Safety and Health, 944 Chestnut Ridge Road, Morgantown, West Virginia 26505 (U.S.A.)

ABSTRACT

Coleman, P.J., 1984. Descriptive epidemiology in job injury surveillance. *Journal of Occupational Accidents*, 6: 135—146.

Descriptive epidemiology is a well-defined set of methods for describing the distribution of diseases among human populations. It is less commonly used in describing accident and injury occurrences, particularly for occupational injuries. In particular, surveillance of occupational injuries in the National Institute for Occupational Safety and Health, Division of Safety Research, is currently carried out using traditional labor statistics and administrative statistics. By combining these data sources with employment figures, risk assessment using epidemiologic methods is carried out where possible.

The data sources are described and their advantages and limitations discussed. The various purposes of surveillance activities are outlined, and the data sources are examined as to how well they serve these purposes. Recommendations are included for improving existing data sources and for planning new data collection efforts so that safety problems in the workplace can be more scientifically assessed and defined.

INTRODUCTION

Work-related injuries in the United States have been described and characterized for a number of years by several different organizations, including the U.S. Department of Labor's Bureau of Labor Statistics (BLS), the National Center for Health Statistics (NCHS), the National Institute for Occupational Safety and Health (NIOSH), and the National Safety Council (NSC), a non-profit, non-governmental organization. Job injury statistics from these sources have provided researchers and others with basic facts and estimates of the numbers and extent of job injuries in the United States. Such information provides a basis for much of the epidemiologic, ergonomic, and technological research in NIOSH's Division of Safety Research. This report provides a brief description of these data systems, a discussion of the role they play in research in the Division of Safety Research of NIOSH, and an evaluation of these sources with recommendations for future development.

SURVEILLANCE AS AN EPIDEMIOLOGIC ACTIVITY

Surveillance as a public health activity has always implied a gathering of and sensitivity to information. Surveillance as early as that practiced by William Farr, Superintendent of the Statistical Department of the General Register Office of England and Wales in the mid-1800s, involved the careful collection, analysis, and interpretation of death records (Langmuir, 1976). Later uses of the term implied a "watching of the exposed", particularly those persons who had known contact with serious infectious diseases such as plague or smallpox. Most recently, the term surveillance has been broadened to include the monitoring of medical care services.

Job injury surveillance as an organized epidemiologic activity is a relative newcomer to both traditional disease surveillance and occupational safety. Aggregate measures of job safety experience have been used for years, but have seldom been used to determine corrective measures. As an essentially control-intensive field, occupational safety has traditionally used injury statistics to monitor global trends, and to call public attention to safety problems. State and federal government agencies have used statistics administratively to monitor activities and to evaluate programs. By contrast, small to medium workplaces have few injury or illness cases to base analysis on, and may choose not to use national or state statistics for internal purposes. At this local level, workplace safety decisions about which controls are necessary, or what should be done about a particular safety problem, are more often based on single accident investigations. Injury statistics have not generally been used in such day-to-day decision making.

As a consequence, existing data systems have been slow to adapt to the more scientific demands of users who want more than just frequency-of-occurrence data. Repeated attempts to analyze small numbers of cases, or to exploit administrative data for causal relations, have frustrated researchers and others seeking information for more basic safety measures and risk factors. While there has been little change in the most broadly-based data sources on occupational injuries in the United States in recent years, there has been an increasing awareness that safety measure effectiveness is critically dependent upon accurate information. With the realization that epidemiologic methods offer means for the scientific assessment of injury problems as well as diseases, occupational safety and health researchers have begun to bring together the necessary pieces of the picture to maximize the usefulness of existing data and make recommendations for filling the remaining gaps.

INJURY SURVEILLANCE AND ITS PURPOSES

Surveillance in the occupational injury arena has traditionally not involved a close monitoring of the exposed, since the great majority of workers in all industries are at some risk of accidental injury. Instead, as with death certificate analysis, the basis for injury surveillance in the past has been reports of injuries after they have occurred. These case series of reports (worker's compensation first reports of injury, OSHA Log entries describing occupational injuries or illnesses) have traditionally contained data on the immediate circumstances surrounding the accident and injury, with few details that would allow inferences to be drawn concerning exposures or background factors.

Occupational injury surveillance from a government viewpoint is therefore best understood as a necessary precursor to other more focussed and specialized activities, equally important to the safety and injury prevention process. In the Division of Safety Research, these include engineering and ergonomics research, and document development and information dissemination activities. While incoming data are critical to effective surveillance, the uses of processed information derived therefrom are equally critical. Functionally, surveillance activities serve three essentially distinct purposes:

(1) to collect and interpret organized data which can be fed back immediately to the public to inform and motivate to action when hazards are clearly present and acceptable countermeasures are available;

(2) to produce analyses of exposure/injury data, including cost-effectiveness arguments, which can provide rationale, justification, and priority setting for other legislated programs, such as standards enforcement; and

(3) to provide indications of elevated risk among high-risk groups which lead to hypotheses for further study and research.

In general, information based on available data alone, often collected to satisfy administrative or legal requirements, may be used effectively to inform the public (as in (1) above), but often falls short when used in cost—benefit calculations and priority setting, and in identifying high-risk areas of work and offering plausible hypotheses as to causes.

SURVEILLANCE IN THE DIVISION OF SAFETY RESEARCH (DSR)

For purposes of conducting research, DSR has utilized a variety of available data to set priorities, detect trends, and provide background for specific research projects and reports. To assess the usefulness of each, several basic criteria were set out which defined strengths and weaknesses for various purposes. The criteria used included the following:

(1) Timeliness. Some surveys present data that are two or more years old, and it is sometimes difficult to judge whether conditions have changed in the intervening period.

(2) Scope. A survey may collect and present data by levels of industry but not occupation, or on a national level but not by region or state. Such aspects determine its usefulness for surveillance.

(3) Degree of detail available. A national survey data source may publish

only summarized tables of injury count by industry, but might not collect or make available case-by-case data.

(4) Inclusion of both numerator and denominator data. Since true risk assessment requires incidence rate data (injury-count numerators and exposure-measure denominators), data sources with numerator-only data are limited in their surveillance potential.

(5) Uniformity of data inclusion. Because of legal and administrative variations in the definitions of compensable injuries, some data sources do not permit cases to be aggregated for certain levels of analysis.

Table 1 summarizes some of the main characteristics of these data bases, and the strengths and weaknesses vis-à-vis the criteria listed. Each data source is further described with examples in the paragraphs following.

TABLE 1

Characteristics of data systems used in Division of Safety Research Surveillance

Data system	Published when and in what form	Geographic specificity	Time between last date covered and date available	Major strengths	Major deficiencies
NIOSH-CPSC's National Electronic Injury Surveillance System	Case-file of emergency-room-treated occupational injuries. Published internally only	Provides national estimates of emergency-room-treated injuries	Daily printout of cases occurring in previous 8 days. Computer data base updated monthly	Timely case data on emergency-room-treated injuries	No occupation or industry coded
BLS Annual Survey of Occupational Injuries and Illnesses	Annual report of incidence rates by industry	Provides national estimates for the survey year	News release on overall measures at 11 months. Full report — 16 months	Accurate measures of industry incidence rates.	Time delay; limited to industry
BLS Supplementary Data System	Case file of workers' compensation cases by state (magnetic tape, summary tables)	Provides state estimates of workers compensation cases	Magnetic tape covering a calendar year is available 16—24 months after end of survey year	Detailed case characteristics by occupation and industry	Time delay. No exposure measures. State to state variations
National Center for Health Statistics Health Interview Survey	Periodic Report of numbers and incidence rates by age group, various conditions	Provides national estimates for the survey year	Full report available — 16 months later	Accurate measures of individual injury experience	Time delay. No occupation or industry coded
National Safety Council Accident Facts Publication	Annual report of numbers, incidence rates, costs	Provides national and state estimate for some measures	Preliminary condensed edition — 3 months. Full report — 9 months	Comprehensive counts of cases by industry, state. Summarize many systems	

National electronic injury surveillance system (NEISS)

Hospital emergency rooms treated 36% of injuries that occurred at jobs or businesses in 1975 (Ries, 1978). Assuming this proportion would yield a fairly representative sample of occupational injuries, the Division of Safety Research in 1981 entered an agreement with the originator of the system, the U.S. Consumer Product Safety Commission, to obtain daily reports of job injuries from a random sample of hospitals across the nation.

These hospital emergency room reports provide case-series data on individual episodes of injury, when these are indicated to be work-related by the injured party. Specific data items include age, sex, diagnosis (nature of injury), locale, part of body injured, type of accident or exposure, source of injury, time and date of accident and treatment, severity, hospital where treated, and a short narrative comment describing what happened.

Daily computer printouts of the previous day's collected reports are given to the Safety Surveillance Branch. When received, most reports have treatment dates of two to five days earlier, while others are collected up to several weeks later. In addition to the daily listing of cases, a long-term mass storage data base is updated with recent cases every two weeks.

Standard frequency distributions of accident and injury characteristics, including accident type and injury diagnosis, are produced weekly, monthly, and annually. The narratives are scanned weekly to detect unusual patterns, accident types, sources of injury, and unexpected exposures.

Table 2 shows incidence rates for job injuries as determined by NEISS data. Rates were estimated by using NEISS-reported injury estimates for the nation as the numerators, while denominators were taken from national employment surveys that are updated monthly. As can be seen, injury incidence rates vary significantly by age, and sex. While it is probable that these patterns are consistent over the different types of injuries represented,

TABLE 2

Estimated incidence of occupational injuries treated in 66 hospital emergency rooms, by sex and age — United States, January 1, 1982—December 31, 1982

Age group	Males		Females	
	Incidence	Rates[a]	Incidence	Rates[a]
16—17	58,100	8.2	16,900	3.0
18—19	201,500	12.0	55,500	4.2
20—24	585,900	8.1	165,000	2.8
25—34	840,400	4.7	240,600	2.1
35—44	378,600	2.9	126,700	1.5
45—54	209,400	2.1	94,200	1.6
55—59	82,100	1.8	37,300	1.5
60—64	14,800	1.0	6,900	0.9
Total	2,419,900	4.0	762,500	2.0

[a]Per 100 workers/year. These rates estimate injuries per 200,000 hours worked, the equivalent of 100 workers per year. A typical full-time worker is estimated to work 2,000 hours in a year.

these and other tests still remain to be done. Of particular interest will be tests of the relatively large male to female risk ratios exhibited by the younger age groups when nature of injury is held constant. Other hypotheses suggested by the data are that part-time work poses higher risk, and that high-risk industries such as construction have peak employment in the summer and autumn months.

Annual survey of occupational injuries and illnesses by industry

Since 1971, the U.S. Department of Labor — Bureau of Labor Statistics has collaborated with state agencies to survey workplaces for occupational injury and illness occurrences. The series of annual reports presenting the survey results represents the single most comprehensive cataloging of the occupational injury incidence rates by industry available today. As an index of risk for industry categories (2-, 3-, and 4-digit Standard Industrial Classification), the Annual Survey is widely quoted as a measure of effectiveness of federal government occupational safety programs, including the Occupational Safety and Health Administration's regulation of workplace safety. Examples of widely quoted results from this survey include the overall industry incidence rate for the U.S., which was 8.3 injuries and illnesses per 100 full-time workers in 1981, and the total number of job-related injuries, which was 5.3 million in 1981.

Researchers have found the Annual Survey useful for identifying high-risk industries, for ranking industries by various measures of safety performance, and for associating time trends in industry incidence rates with known state or federal programs. Measures published in addition to the injury and illness incidence rate include the lost workday case incidence rate (3.9 lost workday cases per 100 full-time workers in 1981), and the lost workday incidence rate (60.4 lost workdays per 100 full-time workers in 1981).

Supplementary data system (SDS)

A system of records increasingly utilized by job safety researchers is that produced by the U.S. Department of Labor — Bureau of Labor Statistics, entitled the Supplementary Data System (SDS). Based on uniformly coded workers' compensation claims from 33 states, this data system is supplementary to BLS's earlier-established Annual Survey of Occupational Injuries and Illnesses. The latter produces injury incidence rates by industry, but provides no case-by-case detail nor any statistical information on the circumstances surrounding accidents. The SDS case-files provide data similar to that provided by the NEISS files. In addition to age, sex, nature of injury, part of body injured, type of accident or exposure, source of injury, and time and date of accident, occupation and industry are also coded. These two items allow some comparison of SDS and Annual Survey data, and have been used, along with appropriately chosen employment data, to develop occupation-specific injury incidence rates.

Table 3 below presents selected occupations and their injury ratio indexes (Root et al., 1981). While denominator data for the SDS injuries in the form of state employment estimates present some statistical difficulty, the ratio index overcomes this by measuring risk in an occupation relative to the risk for the entire industry.

TABLE 3

Selected occupational injury ratio indexes[a]

Occupation	Weighted percent injuries	Percent employment	Ratio index
Highest 5:			
Warehouse laborers, N.E.C.	1.99	0.20	9.95
All other laborers	10.80	1.90	5.68
Structural metal craftworkers	0.39	0.11	3.55
Roofers, slaters	0.41	0.20	2.80
Sheetmetal workers and apprentices	0.56	0.20	2.80
Lowest 5:			
Restaurant, bar managers	0.29	0.60	0.48
Engineering and science technicians	0.38	1.15	0.33
All other managers and administrators	2.14	9.32	0.23
All other clerical and kindred workers	3.07	18.01	0.17
All other professional, technical and kindred workers	1.46	8.81	0.17

[a]Ratio index is weighted percent injuries over percent employment. Percent injuries estimates derived from Supplementary Data System files for 1978. Percent employment estimates derived from BLS National Industry Occupational-Employment Matrix, 1978.

National health interview survey

Among occupational safety and health practitioners, a less well-known but nevertheless valuable survey is the National Health Interview Survey, which is conducted annually by the National Center for Health Statistics among the U.S. civilian non-institutionalized population. Its usefulness to the public health professional and scientist is unquestioned due to the large number of households (42,000 in 1979) surveyed, and to the extensive information collected on the health status of individuals (111,000 from the 42,000 households). Perhaps because of the small number of questions relating to injuries on the job, the Health interview Survey has received less attention from safety professionals and researchers. Its value to DSR lies in its national scope, its ability to provide injury incidence rates, and its reputation as a carefully designed scientific assessment of the population's injury and illness experience. Table 4 presents relevant survey results on job injuries.

TABLE 4

Selected injury characteristics[a], United States, 1975.
(Number of episodes in thousands)

Selected characteristic	All episodes	At job or business	Not at job or business	Unknown
All episodes	48,256	11,411	34,409	2,435
Sex				
Male	23,843	8,796	14,312	735
Female	24,413	2,615	20,097	1,700
Age				
17—44 years	32,757	8,828	22,323	1,606
45—64 years	10,796	2,364	7,887	545[b]
65 years and over	4,703	220[b]	4,199	284[b]
Medical attention				
Attended at emergency room	15,444	4,150	10,964	330[b]
Attended, but not at emergency room	19,687	5,168	13,135	1,384
Attended, place unknown	1,875	435[b]	1,262	178[b]
Not medically attended	11,250	1,659	9,048	543[b]

[a] From the National Health Interview Survey, 1975.
[b] The relative standard error is more than 30 percent: estimates given solely for combining with other cells.

National Safety Council — Accident Facts

The National Safety Council's publication each year of Accident Facts, a summary of information from many sources on accidents of all kinds, is an event anticipated by safety professionals and safety researchers as well as newspaper reporters and others interested in safety. The publication includes estimates of fatal occupational injuries, the frequency, costs, and characteristics of disabling occupational injuries, and historical information on trends in death and injury rates in various industries. Figures from Accident Facts are widely quoted, and are used to provide background information on project plans, comparative evaluations of industries, and to assess trends for effective administration of research programs.

ADVANTAGES AND LIMITATIONS

As is evident from the above, data sources available provide broad indicators of risk, often in the form of numbers of injuries, sometimes in the form of incidence rates. Each of the sources described above provides detail on different aspects of injuries, accidents, and workers exposed. As starting points for research projects, or for some of the other uses outlined such as informing the public, these sources are invaluable. In terms of providing all of the data necessary to determine a potential countermeasure, to identify causes in terms of risk factors, or even to identify risk differentials in a population of workers, these data systems achieve maximum usefulness only as a first step toward problem solution. Analyses of such data bases cannot substitute for, and should not be confused with, the results of carefully defined and skilfully executed studies.

With respect to the three major functions of surveillance of job injuries, existing data sources vary in their usefullness. Existing data are perceived as adequate for public attention purposes, partly because few demands are made on their scientific or statistical validity. This is not to say that published statistics on job injuries are invalid; indeed, the Bureau of Labor Statistics and the National Center for Health Statistics have well-deserved reputations for a high degree of quality and validity. Rather, such statistics are more often used in the popular press to draw attention to a high-risk industry, or to alert the public to working conditions that deserve improvement. In this sense, they are useful in identifying areas that need closer scrutiny, but this process is not one which demands absolute rigor. Instead, the function of informing is completed once the information is disseminated, and government disseminators can assume that the public will use such statistics for a wide variety of purposes, including the use of private resources to improve prevention. In particular, data from the NEISS, Supplementary Data System, the Health Interview Survey, and the National Safety Council serve important purposes in providing broad public information on our accident and injury experience as a nation.

Other administrative uses of surveillance data are clearly more demanding on rigor and methods of analysis. If other government agencies, for example, are to use published surveillance data for problem identification or priority setting, they need information on the populations of workers exposed to specific risks and conditions. Ideally, they need information which indicates probable causes, rather than simply an indication of where problems exist. Numerator-only data such as case-series of injury reports cannot provide information on exposures or populations at risk.

An exception to this is the identification of hazardous products. When a poorly designed product is introduced to widespread use, a case-series surveillance system such as the NEISS can detect time-specific indications that the product is involved in increasing numbers of injuries. The implied denominator in such a situation is a time period, and the sudden increase in cases may very well predict a non-random factor such as the product or behavior surrounding its use.

Unfortunately, while this approach has worked well for consumer product injuries, occupational injuries rarely display such clear-cut time trends. Rather, chronic or long-standing conditions seem to characterize occupational injury risk, and these demand risk data ordered by condition instead of time period.

Most demanding are the scientific and research needs of users concerned with the underlying environmental, social, behavioral, and physical determinants of accidents and injuries. The disadvantages of the data systems discussed above in this area are rather easily stated. Where the SDS data base can provide clusters of occupations, sources of injury, and accident types, as pointers to potential problems, it cannot by itself distinguish worker or management behaviors from physical factors as risk determinants. The NEISS case-series is similarly limited, in that details on pre-existing conditions are not included. The Health Interview Survey provides incidence rate data, but few inferences can be drawn regarding risk factors; detail is insufficient to pinpoint any but the most general patterns of injury risk among U.S. workers.

A particularly significant limitation, shared by all of the above data systems, is that none is adequate to provide meaningful appraisals of the effectiveness of various preventive measures. With the exception of the NEISS system, which can reflect a resultant decrease in a product's involvement in injuries, the surveillance tools discussed can neither evaluate the impact of specific government programs, nor the benefits of a particular safety improvement introduced into a workplace. It should be clear that all of the purposes to which established survey results are put would be better served if true risk data on workers, occupations, industries, and suspected hazards were collected.

CONCLUSIONS AND RECOMMENDATIONS

Surveillance of occupational injuries is an activity that is critical to the intelligent and informed selection of priorities for reseach in accident prevention, and for the measurement of effectiveness of existing control measures. Accident investigations and critical incident studies constitute tools and techniques which are critical in their own right, but at a more localized level than surveillance. What has been absent is attention to several gaps in the process, which starts with the initial collection of accident and injury data and ends with a measurable, verifiable reduction in risk or a reduction in loss from accidents and injuries.

These gaps represent steps in the process which have often been perceived as unnecessary or as too costly, perhaps because, in many cases, obvious countermeasures have been apparent and yet not easily implemented. In brief, some of the gaps can be characterized as follows:

(1) Incidence rate data by industry categories provide a broad indication of injury risk, but problems defined by industry practices are often narrow and highly specialized. Once they are defined this way, more general problems which cut across industries are not likely to emerge as significant, any may persist unrecognized.

(2) Risk assessment needs to be done for a variety of controllable variables (amount of training, experience, work practices and behaviors) so that true risk factors can be revealed and better controlled. Too often decisions on countermeasures are made on the basis of numerator data alone.

(3) While research on feasible controls is proceeding, scientifically gathered risk information can often lead to immediate control if disseminated properly.

(4) While the safety literature is rich in anecdotal evidence that popular approaches to accident prevention are effective, little scientific evaluation research is done to confirm this. In particular, surveillance data systems are insensitive to success stories on an establishment level.

Recommendations for occupational injury surveillance data systems based on these gaps are:

(1) Job injury surveillance data must be enhanced to incorporate a wider range of controllable variables. Demographic classifications will continue to be necessary to allow problems of a manageable size to be identified, but general risk factors will allow much more generalizable controls to be implemented. For example, training, experience, and the wearing of personal protective equipment are proper subjects for inclusion in injury reporting systems.

(2) Greater attention to denominator data needs will increase the usefullness of existing surveillance and epidemiologic data. Special studies will still be necessary, but more informed decisions can be made on actual risk data.

(3) Accident investigations and other specialized techniques can benefit from better risk data and more detailed knowledge of risk factors. Better documentation of the decision options generated by investigations, as well as systematic evaluation of the decisions actually taken will help in better appraising the value of these techniques. A better understanding of the respective roles of surveillance, epidemiology, analysis, and control, will also help facilitate this.

REFERENCES

Accident Facts, 1982 Edition. National Safety Council, Chicago, Illinois.

Langmuir, A.D., 1976. William Farr: Founder of modern concepts of surveillance. Int. J. Epidemiology 5(1): 13—18.

Occupational Injuries and Illnesses in the United States by Industry, 1981. U.S. Department of Labor, Bureau of Labor Statistics. Bulletin 423, 1982.

Ries, P., 1978. Episodes of persons injured: United States, 1975. Advance Data 18 (1978), 1—11. National Center for Health Statistics, DHEW Publication No. (PHS) 78-1250.

Root, N. and Sebastian, D., BLS develops measure of job risk by occupation. Monthly Labor Review, October 1981: 26—30.

USE OF CENSUS DATA COMBINED WITH OCCUPATIONAL ACCIDENT DATA

ELISABET BROBERG

National Board of Occupational Safety and Health, S-171 84 Solna (Sweden)

ABSTRACT

Broberg, E., 1984. Use of census data combined with occupational accident data. *Journal of Occupational Accidents*, 6: 147—153.

The number of accidents in different occupations, age groups and sex are compared with the number of persons working in corresponding groups in Sweden. The accident frequency rates show that food processors have the highest rate, especially butchers and meat preparers. The study also shows that in all occupations men have higher rates than women and that the youngest age group has the highest rate. Thus the group with the highest rate is male food processors up to 24 years.

The study shows that the accident frequency rates have decreased from 1965 to 1980 for almost all occupations. One remarkable exception is food processors, where the rate has increased.

INTRODUCTION

For every year it is possible to calculate the risks in different industries. It is however not possible to break down the working hours in those industries into the different occupation-groups that are working there. The rate for an industry is thus a mixture of the risks for the different occupations, both low and high risk groups. It is therefore desirable to calculate the risks in different occupations.

A census of the population is performed every fifth year in Sweden. The latest one was in 1980. This census makes it possible to get information of the economically active population by occupation, industry, sex, age, county etc. (Folk- och bostadsräkningen 1980, 1981).

All economically active persons — employees, employers and self-employed — are compulsorily insured against occupational injuries by the Work Injury Insurance Act. Occupational injuries (both accidents and diseases) are registered in the Information System on Occupational Injuries (abbreviation ISA) at the Swedish National Board of Occupational Safety and Health. A description of ISA is given by Andersson and Lagerlöf (1983). Lists of the variables and their classifications are given by Broberg and Lagerlöf (1983).

The classifications used for occupation and industry are the same both in ISA and the Census, as are the main principles for using the classifications. Consequently, it is possible to combine the ISA data on occupational accidents and the data from the Census and thus to obtain frequencies for the risks in different groups (the accident frequency rates).

RESULTS

There are of course shortcomings in comparisons between accident frequency rates for different occupations. The Census shows for instance that different occupations have different relative shares of the economically active population working whole days and halfdays. Except for certain occupations, however, calculations show that these differences are negligible. Table 1 shows the economically active population in Sweden in 1980 and the accident frequency rates for men and women, i.e. the number of occupational accidents per 1000 economically active persons.

TABLE 1

The economically active population in Sweden, 1980, and the number of occupational accidents per 1000 economically active persons (unweighted and weighted rates)

	Number of persons			Frequencies	
	1—19 h per week	20+ h per week	Total	Unweighted rate	Weighted rate
Men	59 427	2 148 670	2 208 097	41.6	42.2
Women	199 360	1 604 278	1 803 638	12.2	12.9
Total	258 787	3 752 948	4 011 735	28.4	29.3

In the weighted accident frequency rate the number of persons working 1—19 hours per week has been multiplied by a factor of ½. We can see that the unweighted and weighted rates are very much alike.

The problems are more of a practical kind. If you start to split your population by occupation, sex, age, etc. you soon find that many groups are too small to give stable results, even if you start with a population of 4 million individuals.

Occupations with high risks

The occupations with the highest accident frequency rates are presented in Table 2. The mean number of sickness days per accident (severity rate) are also given for the occupations. The table only covers occupations with at least 1000 economically active persons. The rates would otherwise be too unstable.

TABLE 2

The occupations with the highest number of occupational accidents per 1000 economically active persons (accident frequency rates), and the mean number of sickness days per accident (severity rate) for these occupations. Year: 1980

Occupation	Accident frequency rate	Severity rate
Butchers and meat preparers	231	14
Food processors not elsewhere classified	165	19
Metal casters and moulders	154	20
Sawyers	133	24
Metal smelting, converting and refining furnacemen	121	21
Miners and quarrymen	121	25
Metal processors not elsewhere classified	111	20
Insulators	108	23
Metal rolling-mill workers	103	24
Sheet-metal workers	102	18
Stevedores	102	25

The occupation with the highest accident frequency rate is butchers and meat preparers with 231 accidents per 1000 economically active persons. They have a severity rate of 14 sickness days per accident which is one of the lowest.

The occupations with the highest accident frequency rates include several in the area of basic metal industries. These are metal casters and moulders with an accident frequency rate of 154, metal smelting, converting and refining furnacemen 121, unspecified metal processors 111, and metal rolling-mill workers with 103 accidents per 1000 workers.

Sawyers (133) and miners and quarrymen (121) also have high rates.

A division of the population by sex gives about the same occupations with high rates for both men and women but with higher rates for the men as the women have lower rates for all occupations.

Food processors by sex and industry branch

A division of the population by occupation, sex and branch of industry is possible if the population is sufficiently large.

The group with the highest frequency rate is food processors. They work mainly in two industries, the food industry and the wholesale and retail trade. The risks in the different areas are given in Table 3. Male food processers in the food industry have a much higher frequency rate, 151, than the other groups. In spite of that, the two male groups have about the same accident pattern, i.e. 50% of the accidents are injuries caused by pressing, cutting, etc. with a handled object or instrument, mainly knives. They also have about the same number of accidents caused by objects, machine parts, etc. in motion or overexertion of a part of the body. The female groups have also the same type of accidents but the injuries caused by handled objects or instruments are 25—30% of all accidents. They also have a lot of falls (20—25%).

TABLE 3

Number of occupational accidents per 1000 working food processors in Sweden by sex and branch of industry. Year: 1980

Branch of industry	Men	Women	Total
Food industry	151	65	120
Wholesale and retail trade	62	62	62
Average	135	65	112

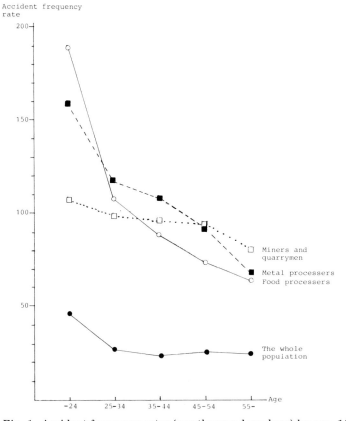

Fig. 1. Accident frequency rates (per thousand workers) by age, 1980. The whole population and the three occupations with the highest rates.

Risks by occupation and age

It is possible to split the economically active population by occupation and age simultaneously. This makes it possible to calculate accident frequency rates in different age groups. Figure 1 shows the rates for different age groups for the whole population and for the three occupations with the highest rates. We can see that within every occupation the youngest group has a much higher rate than the older groups.

The differences between the age groups are however quite different between the various occupations. The occupation with the highest rate, food processors, has a very high rate in the youngest age group, much depending on their special working conditions. Miners and quarrymen, who do similar work in all age groups, are also subject to similar risks.

The risks by occupation, age and sex

If the population is sufficiently large a division is possible by occupation, age and sex simultaneously. In Table 4, the population is divided by sex and age for the whole population and the two occupations with the highest frequency rates. The rates for men follow the pattern for the whole group, i.e. the youngest age group has the highest rate and the rate decreases with age. The rates for women are more irregular. The youngest group has the highest rate but the patterns for the later age groups are different for different occupations. Women always have a lower rate than men of corresponding age.

TABLE 4

Number of accidents per 1000 economically active persons by age and sex for the whole population, food processors and metal processors

Age	The whole population		Food processors		Metal processors	
	Men	Women	Men	Women	Men	Women
−24	71	18	238	86	166	79
25−34	42	8	131	53	125	53
35−44	35	10	104	55	112	77
45−54	37	13	82	60	98	51
55+	33	15	65	61	68	53
Average	42	12	135	65	114	63
No. of accidents	91933	21989	3588	832	3263	207

Comparison with earlier censuses

A comparison is made with the frequency rates from years when earlier censuses were performed. A decrease of the rate from 39 to 28 between 1965 and 1980 is notified for the whole population. Most of the occupations have a decrease in the rate, at least from the year 1970. One exception is food processors. Their rate has increased from 83 to 114 between 1965 and 1980.

In Fig. 2, the development from 1965 to 1980 is shown for the whole population and for the three occupations with the highest accident frequency rates in 1980.

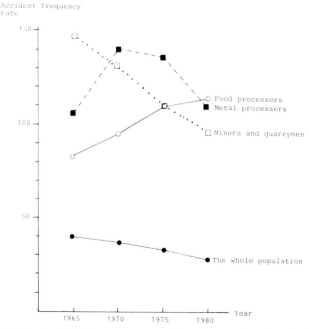

Fig. 2. Accident frequency rates (per thousand workers) 1965—1980. The whole population and the three occupations with the highest rates.

DISCUSSION

These are just some examples — the most important ones — of the results which it is possible to get when combining census data with occupational accident data. The possibility of isolating groups with high accident frequency rates increases when accident data are combined with population data from the Census. This depends mainly on the possibility of obtaining a breakdown of the working population by occupation, as the annual information of working hours per branch of industry, which is available in Sweden, is a mixture of all occupations working in one industry.

The additional possibility of calculating the accident frequency rate by sex and age in the different occupations is of course a further improvement. The main restriction is that the groups soon become too small to get stable results and thus it is not possible to split the population into all the groups which one would like to study.

REFERENCES

Andersson, R. and Lagerlöf, E., 1983. Accident data in the new Swedish Information System on Occupational Injuries. Ergonomics, 26: 33—42.

Arbetsskador, 1980 — yrkesrelaterade risker, 1983. (Occupational Injuries 1980 — Occupational-Related Risks). Sveriges officiella statistik, Arbetarskyddsstyrelsen (National Board of Occupational Safety and Health), Statistiska centralbyrån (Statistics Sweden), Sweden. (Summary in English)

Broberg, E. and Lagerlöf, E., 1983. A short description of the Swedish Information System on Occupational Injuries (ISA). Arbetarskyddsstyrelsen (National Board of Occupational Safety and Health), Sweden.

Folk- och bostadsräkningen 1980, 1981. (Population and Housing Census 1980). Sveriges officiella statistik, Statistiska centralbyrån (National Central Bureau of Statistics), Sweden. (Summary in English).

HAND INJURIES IN SWEDEN IN 1980

ANNIKA CARLSSON

The National Board of Occupational Safety and Health, S-171 84 Solna (Sweden)

ABSTRACT

Carlsson, A., 1984. Hand injuries in Sweden in 1980. *Journal of Occupational Accidents*, 6: 155—165.

Hands and fingers are injured in 36% of all occupational accidents in Sweden. By using the Swedish Information System on Occupational Injuries (ISA) a survey of the hand injuries is made which shows the most affected branches, occupations, machines and equipment. The most frequent chains of events leading to hand injuries are shown.
  The occupations with highest frequencies of hand injuries are butchers, sawyers and metal casters and moulders.
  Machines, vehicles, lifting and surveyor devices only cause one third of the accidents. The rest are caused by simple equipment like hand tools and construction material.
  Long and complicated chains of events immediately preceding accidents are rare. Loss of personal control, usually loss of hold, precedes half of the injuries. Technical events like ruptures, collapses, machines not functioning, etc. only happen in 15% of the accidents. Amputations are mainly caused by metal and wood processing machines. More than one third of the amputations are preceded by loss of personal control.

INTRODUCTION

The hand is the part of the body that is most frequently injured in occupational accidents. In Sweden, about 36% of all occupational accidents leading to absence from work are hand injuries (Statistics Sweden, 1983). In 1980, 38,038 hand injuries occurred causing about 750,000 days of absence, that is 19.3 compensation days per accident, or about the same as the national average for all occupational accidents.
  The great number of hand injuries has initiated quite a lot of projects and discussions on how to reduce it. But what are hand injuries? Are there typical chains of events leading to hand injuries? The aim of this report is to present a survey of hand injuries to serve as a basis for priority decisions and a background for more detailed studies.

METHOD

The information in the report is taken from injury forms for occupational accidents sent to the social insurance offices and registered in the Swedish

Information System on Occupational Injuries (ISA). All accidents where the injured person was away from work at least one day are supposed to be reported. For every accident at least 37 different variables are registered, e.g., branch of industry, firm, county, nature of injury and bodily location, employment and pay conditions, work schedule, age, sex and occupation of the injured person, etc.

The accidents are described by a chain of events, temporal sequence and the external agencies that are involved in the accident (see Fig. 1). The chain consists of the injury event, contact event and pre-events. For more details about the ISA system see Andersson and Lagerlöf (1983). Unless otherwise stated, the information in this report concerns only employees.

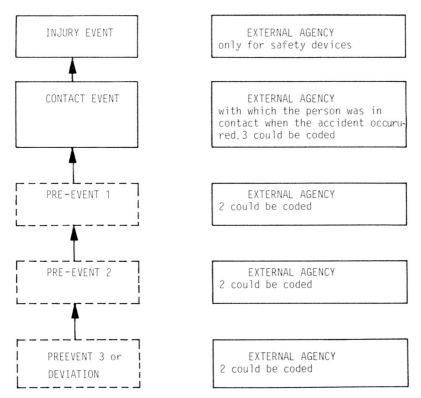

Fig. 1. Classification by chain of events in the ISA system.

RESULTS

In which branches of industry and occupations do people most often get hand injuries?

For all occupational accidents the highest frequency rates of hand injuries are found in manufacturing, the construction industry and mining. The most affected industries are slaughtering (87 hand injuries per one million manhours) and meat preparing (46 hand injuries per one million manhours). The high frequency rates of these industries can be explained partly by the fact that the workers have to stay home for hygienic reasons also when they receive minor injuries. However, the frequency rates of injuries with long periods of absence, and amputations, are also high for these industries.

Other industries with high frequency rates are steel foundries, glaziers and platers in construction work. Many industries have about 20 hand injuries per one million manhours, for example different manufacturers of metal products, vehicles, wood and wood products, shipyards and fish-industries.

The average frequency rate for all industries is 7.5 hand injuries per one million manhours. If you look at occupations instead of industries the tendencies are still more obvious (Table 1). Butchers and meat preparers dominate with 173 hand injuries per 1,000 employees. Sawyers as well as metal casters and moulders have 58 hand injuries per 1,000 employees and glassformers 55 hand injuries per 1,000 employees.

TABLE 1

Occupations with the highest frequencies of hand injuries, 1980 (employees and self-employed)

Occupation	Number of hand injuries	Number of hand injuries per 1,000 employees	Number of immediate amputations
Butchers and meat-preparers	1,796	173	13
Sawyers	746	58	38
Metal casters and moulders	277	58	4
Glassformers	47	55	—
Mechanics	5,149	51	82
Tanners, fellmongers, peltdressers	44	51	—
Joiners, cabinet-workers etc.	1,880	49	69
Glaziers	122	41	—
Sheet-metal workers	1,116	40	13
Metal rolling-mill workers	129	40	2
Total	41,065	10	741

Mechanics received the highest number of hand injuries (5,149 injuries, 51 injuries per 1,000 employees). Other occupations with many injuries but lower frequencies are fitters (3,578 injuries) welders (1,204 injuries) housekeeping service workers (997 injuries) and salesmen and shop-assistants (962 injuries).

Which tools and machines cause most hand injuries?

9% of all hand injuries (3,534) are caused by knives, usually handling accidents of which most are caused by loss of hold (Table 2). Butchers and meat preparers have almost 40% of all accidents with knives. Hammers, etc. cause 953 injuries. Hand tools cause a total of 6,232 hand injuries.

The most frequently reported machine is the slicing machine for meat, bread, etc. (585 injuries). Next come drilling machines, reamers, and tappers (429 injuries). Hand-held machines are reported in 1,674 cases and fixed

TABLE 2

Principal external agencies affecting hand injuries, 1980.
(employees)

Principal external agency	No. of hand injuries	No. of amputations	% hand injuries of all injuries with the agency
Hand-held machines	1674	23	42
Hand tools	6232	8	68
knife	3534	4	90
hammer	953	—	63
Lifting or conveyor device	1428	46	46
Vehicles, motorized implement	2164	39	25
Machines, stationary or mobile	8813	368	75
wood-conversion or woodworking machines	2002	141	84
machines for production of metal or metalware	3333	116	69
drilling machines, reamers, tappers	429	11	80
machines for production or processing of foodstuffs etc.	1375	44	88
slicing machines for bread, meat etc.	585	3	98
Parts of building, etc. or other structure, fixtures or equipment and furnishing, scaffolding, ladders, ground	4061	26	19
Door, window	746	8	57
Floor, etc.	555	6	14
Materials, goods packing, building and construction units (not fixed)	8452	22	33
Metal plate, wire, tube	1360	5	53
Building and constructions units of steel	1085	2	43
Glass	433	1	82
Box, basket, etc.	477	—	20
Total	39038	568	36

machines in 8,813 cases, of which most are caused by wood working machines and machines for metal production and processing.

Other common external agencies are metal tubes, wire, plates, etc. and steel construction units which usually cause injuries by pinching and cutting.

The most serious accidents, leading to amputation of hands or fingers or long period of absence from work, are mainly caused by machines for manufacturing of wood and metal, especially circular saws for splitting with manual feeding, cutters, dovetailing and tenoining machines and surface planing machines. About 30 of the amputations were caused by gloves that got stuck in machines, most often drilling machines.

TABLE 3

Hand injuries — accidents by main event and diseases by suspected cause, 1980 (employees)

Main event	No. of hand injuries	% hand injuries of total no. of injuries of each type
Accidents		
Electrical accidents	94	39
Fire, explosion, blasting	239	32
Fall of person on the same level	2,142	22
Fall of person to lower level	799	15
Contact with chemicals	174	14
Contact with heat or cold	1,267	37
Contact with stationary object	5,405	50
Struck by falling or flying object	2,543	18
Contact with moving machine parts	9,993	76
Blow, kick, etc. from person or animal	558	27
Accident with vehicle	340	12
Overexertion of body part	1,051	7
Accident with handled object	12,913	80
Remaining, unclear	1,520	36
Total accidents	39,038	36
Diseases		
Monotonous or strenuous movements	1,307	15
Chemical substances or products	1,465	34
Vibrations	224	61
Biological factors	191	30
Remaining	267	16
Total diseases	3,454	19
Total accidents and diseases	42,492	34

In which types of accidents are hands injured?

The most common type of accident leading to hand injuries is pressing, cutting, etc. by handled objects or tools (12,913 injuries). Contact with moving machine parts of other moving objects cause 9.993 hand injuries. 80% of all accidents of these types are hand injuries (Table 3).

Hand injuries do not dominate other types of accidents as much. Hands are injured in 18% of the fall accidents (2,941 injuries) and 50% of the accidents caused by blow, press, cutting, etc. against stationary objects (5,405 injuries).

Overexertion of a body part or strenuous movements usually affect other parts of the body. However, 1,051 of the accidents and 1,307 of the occupational diseases affect hands.

The most common cause of occupational hand diseases are chemical substances or products, mainly causing eczema.

What kinds of injury events occur?

Injuries from cutting, tearing and sawing are most frequent (14,152 injuries). Next comes pinching, cut, bite (10,309 injuries), blow (7,538 injuries), piercing, chopping (1,979 injuries) and excessive heating (1,545 injuries) (Fig. 2).

The left hand is more often hurt than the right hand, especially by cutting and blow. Cutting and piercing cause on average shorter periods of absence from work (15 and 10 compensation days, respectively) than blow (26 days) and tearing, twisting (52 compensation days on average).

Which events have preceded hand injuries?

Figure 2 shows that loss of balance or other loss of personal control, most often loss of hold, was reported to have preceded half of the injuries. No preceding event was reported in 8,080 cases.

Due to insufficiently completed forms, the course of events is obscure and an unclear event has been assigned to about 10% of the injuries. The most typical chain of events is cutting by handled object after loss of hold, many of which are accidents with knives in slaughtering. Another common chain of events is injury by a blow after a fall caused by slipping (1,092 injuries) or tripping (454 injuries).

Usually only one preceding event is registered. In those cases where two or more events are registered, the first event is usually loss of personal control. Only 160 hand injuries had three preceding events.

Mechanical events are not so common: If you sum preceding events 1, 2 and 3 the most frequent mechanical events are object coming loose (totally 992 injuries), rupture, distortion (925 injuries), disrupted material flow (697 injuries) and stacking of machine parts, etc. (506 injuries). In total

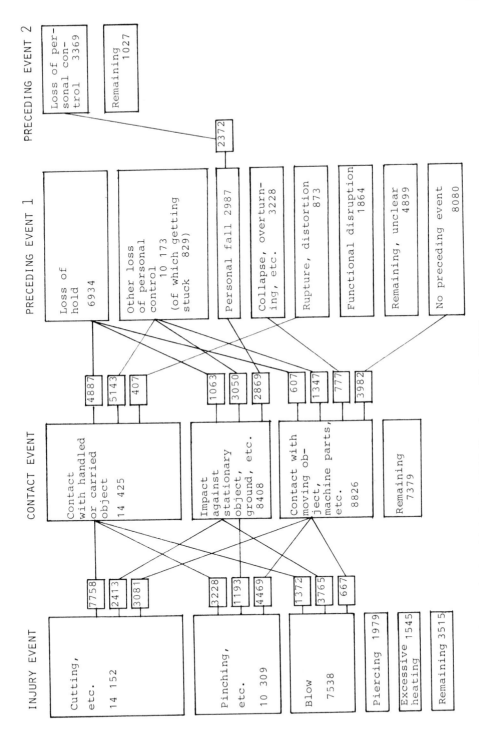

Fig. 2. Chain of events for hand injuries in 1980. Employees ($N = 39038$).

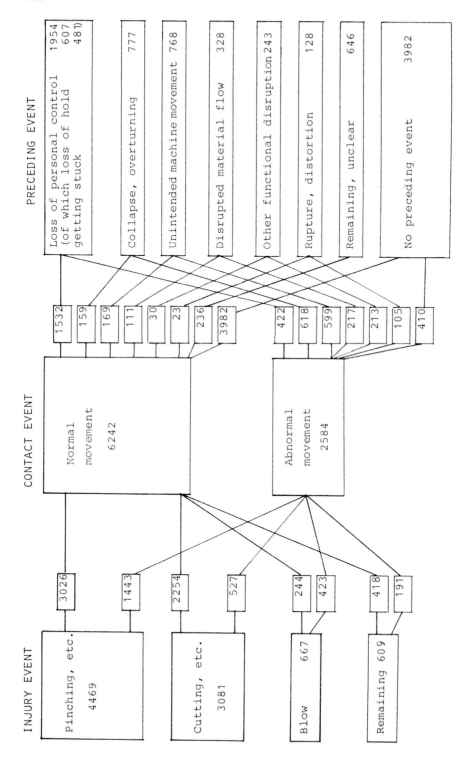

Fig. 3. Chain of events, contact with moving machine parts etc. for hand injuries in 1980. Employees ($N = 8826$).

about 6,000 events of this kind were registered, resulting in about 15% of the injuries.

In more than 500 accidents, gloves were reported to have contributed to the accident by getting stuck or by slipping.

Accidents, where the hands got injured by moving machine parts

Loss of personal control not only dominates accidents with handled or stationary objects but also accidents with moving machine parts (Fig. 3). In more than two thirds of the accidents with moving machine parts (a total of 8,826 injuries) the injuries were caused by normal and intended machine movements. Usually no preceding event was reported or only loss of personal control, and often loss of hold (607 injuries) or getting stuck (481 injuries).

Unintended movements were reported in 768 cases of which about 170 were unintended start. Disrupted material flow was reported in 328 cases, rupture and distortion in 128 cases and other functional disruptions in 243

TABLE 4

Events preceding hand injuries, 1980 (employees)

Principal external agency Preceding event	Slicing machines for bread, meat, etc. Contact with moving object	Cutting wood processing machines	Excentric presses	All hand injuries	Hand injuries leading to immediate loss of part of body
Loss of personal control	36%	24%	8%	51%	38%
• of which loss of hold	23%	11%	1%	18%	7%
• of which getting stuck	0.3%	4%	—	2%	8%
Collapse, overturning, etc.	0.5%	9%	9%	8%	8%
• of which recoil, ejection	—	6%	—	1%	2%
Rupture, distortion	—	0.4%	4%	2%	1%
Unintended machine movement or start	1%	3%	38%	2%	10%
Disrupted material flow	2%	5%	—	1%	2%
Other functional disruptions	0.3%	1%	4%	1%	3%
Remaining, unclear	2%	4%	6%	11%	9%
No preceding event	58%	53%	30%	22%	32%
Total	100%	100%	100%	100%	100%
Total number of injuries	376	1,394	79	39,038	568

cases. Recoil, ejection caused 166 injuries, goods rolling or sliding 112 injuries, objects coming loose 139 injuries and folding 167 injuries.

Of course more complicated machines like excentric presses, strictly regulated by the authorities, have a more complex set of preceding events than, for example, slicing machines for bread and meat (Table 4).

38% of the accidents among excentric presses were caused by unintended machine movements compared to 1% among slicing machines for bread and meat. However, excentric presses also have accidents preceded only by loss of personal control (8% of the accidents with moving machine parts), or with no preceding event (30%).

It is obvious that no specially complicated events are needed to initiate an accident with most machines. Usually a simple loss of personal control is enough.

Accidents leading to immediate loss of part of hand or finger

In total 547 losses of fingers and 21 losses of hands were reported in 1980. This number is in reality too low, since the forms are filled in not later than 14 days after the accident. Severe crushing and pressing injuries may later lead to amputations, which are not covered by the statistics. Injuries by pinching (43%) and cutting (30%) dominate, and most accidents are caused by contact with moving machine parts (71%), usually normal movements (73%). In 32% of the injuries no preceding event was reported and in 38% some kind of loss of personal control.

Most amputation accidents are caused by wood processing machines. The proportions of different types of preceding events are about the same for amputation injuries as for other injuries caused by these machines.

CONCLUSIONS

Hand injuries include too many different kinds of events to enable exhaustive conclusions from an analysis of the whole material. Separate studies on special machines, industries and occupations are necessary. The report gives an idea of how the ISA can be used to analyze different types of accidents and find common patterns that could point out preventive measures.

Some general conclusions can be drawn: The majority of hand injuries are caused by trivial events and trivial equipment. Common tools as knives and hammers still dominate. Only one third of the accidents are caused by machines, vehicles or lifting and conveyor devices.

Usually there is no long complicated chain of events immediately preceding an accident. More than one pre-event is very unusual. Loss of personal control is the most common pre-event, and it is also common that accidents with machines and amputation accidents are only preceded by loss of control.

Only about 15% of all hand injuries are preceded by more technical events like collapse, rupture, fire, unintended machine movements, etc. It could be suggested that these more dramatic events are better reported and less underestimated than trivial events like loss of hold. The event "Material, getting stuck", however, is very commonplace and probably underestimated.

The design of the working places and machinery and the organization of work can make for a severe injury after a simple loss of personal control.

REFERENCES

Andersson, R. and Lagerlöf, E., 1983. Accident data in the new Swedish Information System on Occupational Injuries. Ergonomics 26: 33—42.

Statistics Sweden, 1983. Occupational Injuries Liber Forlag/Allmänna Förlaget, Stockholm.

ESTIMATION OF POTENTIAL SERIOUSNESS OF ACCIDENTS AND NEAR-ACCIDENTS

HEIKKI LAITINEN

Lappeenranta Regional Institute of Occupational Health, PL 175, SF-53101, Lappeenranta 10 (Finland)

ABSTRACT

Laitinen, H., 1984. Estimation of potential seriousness of accidents and near-accidents. *Journal of Occupational Accidents*, 6: 167—174.

The severity of injuries in accidents depends often on chance. Because of this random nature, injuries are a bad measure of the severity of risks for a single factory. Serious injuries are so rare that they normally cannot point to preventive action.

We are able to estimate the potential seriousness of accidents. To study this method the author together with the representatives of a steel factory investigated about 300 accidents and near-accidents. The purpose was to find out the potentially most serious injury and part of the body affected.

The reliability of the estimation and the facts which relate to it will be studied. Also the differences between the estimated risk potential and the risk indicated by real injuries will be studied.

INTRODUCTION

The phases in risk assessment are risk identification and determination of probability and severity of a potential injury (Rowe, 1977). A risk can be identified either before or after a person has been injured.

Traditional inspections and investigations of accidents have been completed by risk analysis and by investigating near-accidents using various methods (Kjellén, 1982; Nill, 1971; Rockwell et al., 1970). Thus it has been possible to increase the material available for risk identification.

One of the main problems is how to distinguish severe risks from slight ones and how to identify severe risks in advance. Most of accident expenses are composed of a small number of serious accidents (Klen, 1981) and they are the most harmful also from the human point of view (Rantanen, 1982).

One method to identify severe risks is the estimation of the potential seriousness of actual accidents. It has been used in some work places (Allison, 1967) but it has been questioned whether its reliability is sufficient.

It has been supposed that workers will especially report very serious near-accidents because of their own subjective risk assessment (Laitinen, 1982). To prove this hypothesis we must have a commensurable measure of seriousness for both accidents and near-accidents.

The only possible method to measure risks identified in different ways seems to be the estimation method. It has been used in risk analysis. Rockwell (1970) has used it in assessing the relative danger of pilot errors while Östberg (1980) used it in various tree felling situations. Both of them found good reliability.

The main purposes of this report are to study the estimation method of potential seriousness of accidents and near-accidents and to test the hypothesis that reported near-accidents are more serious than the actual accidents in a company.

METHODS AND MATERIALS

The material was gathered during a research project at Ovako Oy Steel Factory in Imatra (Laitinen, 1982). A voluntary report system was created for near-accidents and it remained in use also after the research. Investigation groups were formed in each department and they investigated on the spot all actual accidents as well as reported near-accidents. The material gathered related to 223 accidents and 75 near-accidents.

The potential severity of an accident and near-accident was defined as the severest injury possible in that situation (Rockwell, 1970). It was rated by using as a criterion the nature and amount of the burst of energy. Whether the energy could have been directed at people or at a more sensitive part of the body than actually happened was also taken into consideration. This was then called *the potentially injured part of the body*.

All accident and near-accident reports were written in a uniform way for the estimation. In reports the accident sequence was described as completely as possible. The reports were randomized.

A scale was prepared for estimating the potential injury severity (Table 1) and for the potentially injured part of the body. The estimators were asked to indicate the severest injury that could have happened in each case, as well as the potentially injured part of the body. Medical treatment was presumed to be normal in that the injury would not get worse during treatment.

The estimation was made by the works safety officer, the workers' safety representative and the author. At first we went over 25 cases together for practice and in order to make the interpretations consistent. Thereafter each of us estimated the remaining 273 cases. Finally we together dealt with the cases where the estimations differed from each other and found a common solution acceptable to all three.

The reliability of the estimates was calculated on the basis of these three people's estimates of the same cases. The consensus of the part of the body estimates was calculated as the percentage of the pairs of unanimous estimates compared to the total number of pairs of estimates. The reliability of severity estimates was measured by calculating the average of the correlative pairs of estimates.

The validity of estimating an accident's potential severity as agreed upon

TABLE 1

Severity categories of potential injury

Category	Nomenclature	Explanation, examples
1	Small, insignificant	absence less than 3 days
2	Minor	absence 3—30 days; e.g., cuts, contusions, small particles in eye, fracture of fingers or wrist, strain of wrist, burns
3	Significant	absence 1—12 months or permanent degree of disability less than 10%; e.g. fracture of leg or arm, loss of forefinger, loss of finger joint
4	Severe	absence more than 12 months or permanent degree of disability 10—60%; e.g. loss of thumb or two fingers, loss of big toe, loss of whole arm or leg, loss of sight from one eye
5	Critical	death or permanent degree of disability more than 60%; e.g., loss of both arms or legs, loss of sight from both eyes

by the three subjects was tested by using the actual severity (lost working days) as a parallel measure. The estimate is wrong if the actual injury is more severe than the estimate.

RESULTS

Reliability

Agreement on the potentially severest injured part of the body was 73% on average between the three estimators. The correlation stating the agreement of the seriousness estimates was 0.59 on average. In cases where the part of the body estimates was unanimous the agreement of the seriousness estimates was 0.70, which is considerably better.

There was no considerable difference in the part of the body estimates between accidents (72%) and near accidents (78%). On the other hand the potential seriousness was more unanimously estimated in near-accidents ($r = 0.68$) than in the actual accidents ($r = 0.57$).

The agreement between the researcher and the representatives of the factory was not considerably worse in the original categories than the mutual agreement of the representatives of the factory.

Influence of actual injury on reliability

It was assumed that the agreement would rise if the estimates were similar to the actual injuries. The mutual relations between these factors were examined by creating four new, dichotomic variables and studying their partial correlation (Table 2).

TABLE 2

Influence of actual injuries on reliability of estimates, partial correlation matrix, $N = 207$

	1	2	3
1 Body estimate same as injured part of the body (yes/no)	1		
2 Seriousness estimate same as actual seriousness (yes/no)	0.205	1	
3 Body estimate unanimous (yes/no)	0.449	0.067	1
4 Seriousness estimate unanimous (yes/no)	0.140	−0.007	0.243

On the basis of the analysis it seems that the reliability of the seriousness estimates is increased by the agreement on the part of the body estimates, although the influence is not very strong. It seems that the unanimity of the part of the body estimates increases if the estimated injury is directed at the same part of the body as in the actual injury. This correlation is quite strong.

Influence of accident type on reliability

Accident types were formed on the basis of the factor analyses of accidents and near-accidents. The dichotomic variables formed from the categories of the object causing the injury, injury event and the preceding event were used as variables in the analyses.

TABLE 3

Reliability in estimations of three persons according to accident type

Accident type	N	Reliability	
		Seriousness r	Part of the body (%)
Accidental starting	13	0.71	63
Hurting oneself on a stationary object	18	0.69	64
Fall to lower level	10	0.53	89
Falling object	26	0.51	91
Other moving object	38	0.50	68
Flying object	41	0.46	81
Particle in the eye	13	0.39	90
Hurting oneself with a handled object	39	0.36	84
Restrained motion	29	0.34	81
Falling of a handled object	15	0.22	67
Fall on same level	28	0.18	61
Other	3	—	—
Total	273	0.61	73

Estimate reliabilities varied very considerably in the different accident types (variance analysis). The reliability of the seriousness estimate was highest in such accident types as accidental starting of a machine, hurting oneself on a stationary object and falling to a lower level. It was lowest in such types as fall on same level, falling of a handled object and restrained motion (Table 3).

The reliability of the part of the body estimate was the highest in such types as falling object, particle in the eye and falling to lower level. It was lowest in such types as fall on same level, accidental starting of a machine and hurting oneself on a stationary object. The reliabilities of the part of the body and seriousness estimates in different accident types did not correlate with each other.

Influence of the categories of the body on reliability

In potential eye injuries the agreement in estimating the part of the body was best, 91%. The agreement was less than 70% only in the case of the other upper extremities, toes and foot (Table 4).

TABLE 4

Reliability of estimates by part of the body

Potential part of the body injured	Reliability		
	Seriousness* $N = 3 \times 142$ r	Part of the body $N = 3 \times 273$ %	
Head	0.62	72	
Eye	0.68	91	
Fingers	0.72	83	84**
Other upper extremities	0.78	52	
Toes, foot	—	45	79**
Other lower extremities	0.71	70	
Body, back	0.52	78	
Total	0.70	73	82**

*Includes only such cases where all the estimators were unanimous about the part of the body.
**With joint categories.

By combining the categories of finger and other upper extremities, the reliability of the category of the whole arm by using the same estimates turned out to be 84%. Correspondingly, the reliability of the joint category of the whole leg was 79%. Thus the reliability of the group with 5 variables (head, eye, arm, leg, body) turned out to be 82% on average, which can be considered good.

The seriousness of the potential injuries of the extremities was best estimated. The estimates concerning the seriousness of the potential body injuries were the most uncertain. By combining the part of the body categories, they can be made more heterogeneous, but its weakening influence on the reliability of the seriousness estimations remained, however, minor. The reliability of the combined categories was 0.69 while otherwise it was 0.70.

Influence of seriousness categories on reliability

In order to eliminate a mistake caused by coincidental unanimity the category alternatives were compared using Cohen's unanimity coefficient as a measure. In the original categories Cohen's coefficient was 0.39 on average. Unanimity varied from category to category, being the highest in the slightest and in the severest injury category.

Categories 3 and 4 were the most difficult to estimate. One of the reasons for this might have been the fact that their explanations included both the duration of disability and the permanent degree of disability. In some cases these may be contradictory. For example the loss of a finger joint does not necessarily lead to a disability of more than a month.

Decreasing the number of categories considerably increased the reliability (variance analysis). The highest unanimity is achieved with a variable with two categories by leaving either the slightest or the severest category alone and by combining the rest (Table 5). The other category is thus very heterogeneous which means that the advantage of high reliability is lost. The most recommendable variable with two categories seems to be the alternative where categories 1--2 are combined as slight and categories 3--5 as severe accidents.

TABLE 5

Average unanimity of three estimators in seriousness estimations in various combined categories measured by Cohen's coefficient (r_c)

4 Categories	r_c	3 Categories	r_c	2 Categories	r_c
1—2, 3, 4, 5	0.42	1—3, 4, 5	0.47	1, 2—5	0.57
1, 2—3, 4, 5	0.46	1, 2—4, 5	0.53	1—2, 3—5	0.53
1, 2, 3—4, 5	0.43	1, 2, 3—5	0.48	1—3, 4—5	0.50
1, 2, 3, 4—5	0.40	1—2, 3—4, 5	0.46	1—4, 5	0.55
Total	0.43	1, 2—3, 4—5	0.48	Total	0.54
		1—2, 3, 4—5	0.43		
		Total	0.48		

The highest reliability in the combinations of three categories is achieved by keeping the slightest and the severest category independent and by combining categories 2—4. The second category is thus quite heterogeneous. As a whole it is better to keep only the slightest category independent and to combine categories 2—3 as severe and categories 4—5 as critical accidents.

Lost working days and potential severity

When examining the parallel validity of the common seriousness estimate, lost working days caused by the accidents were used as a criterion. The incorrectness of the seriousness estimate is indicated only if the actual injury is more severe than the one estimated as potentially the most severe. This was the case in 8% of the cases. In 35% the potential seriousness was estimated to be the same as the actual one and in 57% larger.

Of the incorrect cases in six the estimation was 2 lost working days at the most as they were 6 on average. In eleven cases the estimate was 30 lost working days at the most as they were 41 on average. Thus the incorrect estimates were minor.

There was no actual severe or critical injury in the material but 22—26% of the actual small, minor and significant accidents were estimated potentially severe or critical. In this respect, the actual severity had no significant effect on the potential severity (chi-square test).

Potential severity of near-accidents

The potential severity of near-accidents was significantly larger than that of accidents (Table 6). More than half of the near-accidents were critical, which means that there was a possibility of death or a very serious permanent disability. Only 10% of accidents were potentially so severe. Most of the near-accidents were of a serious type — falling, flying and other moving object. However, the near-accidents were also more severe than accidents, when the accident type was held constant (log-linear models).

TABLE 6

Potential severity of near accident and accidents

Potential severity	Near-accidents	Accidents
1 Small	2	22
2 Minor	9	79
3 Significant	10	65
4 Severe	12	32
5 Critical	42	25
	75	223

$x^2 = 69.80$
$p < 0.001$

DISCUSSION

The reliability in estimate of potential seriousness was reasonable, but not so good as in Rockwell's (1970) and Östberg's (1980) studies. Unlike this study, they had a limited number of errors or situations, but a lot of estimators.

The estimation method seems to be worth further development. Reliability might be improved, e.g. by developing categories and instructions. The method should be tested with different material and numerous subjects.

One of the advantages of the method in research would be the possibility to measure commensurably the risks identified by different methods (e.g. accidents, near-accidents, inspections and risk analysis). Classification by severity is possible with less material than with measuring methods based on actual injuries and this saves research costs.

About 25% of the accidents were estimated potentially severe or critical. Thus, estimating actual accidents could be useful in routine use.

I agree with Allison (1967) that severe risks can be identified and preventive actions directed with the method. It might also help in activating the safety work in the whole organization. The demands of reliability in practical use are not so great as in research.

More than a half of the reported near-accidents was estimated critical. This supports the hypothesis that workers are prone to report especially severe near-accidents. The result seems to emphasize the usefulness of a reporting system in identifying the risks of a workplace.

REFERENCES

Allison, W.W., 1967. How to foresee tragic accidents by use of The High Potential Accident-prone Situation Hazard Control method. Transactions of the National Safety Congress 2: 4—8.

Kjellén, U., 1982. An evaluation of safety information systems at six medium-sized and large firms. Journal of Occupational Accidents 3: 273—288.

Klen, T., 1981. Economic losses due to occupational accidents in forestry. Research 176. Institute of Occupational Health, Helsinki, 174 pp., (in Finnish, with English summary).

Laitinen, H., 1982. Reporting noninjury accidents: a tool in accident prevention. Journal of Occupational Accidents 4 (2—4): 275—280.

Nill, E., 1971. Schadenkontrolle, Durchbruch zur integrierten Arbeitssicherheit. Sicherheitsingenieur 2: 8—13 and 3: 9—14.

Rantanen, J., 1982. Effect of accidents on public health and national economy. Journal of Occupational Accidents 4 (2—4): 195—203.

Rockwell, T.H., Bhise, V.D., Clevinger, T.R., 1970. Development and application of a non-accident measure of flying safety performance. Journal of Safety Research 4: 240—250.

Rowe, W.D., 1977. An Anatomy of Risk. Wiley & Sons, New York, 483 pp.

Östberg, O., 1980. Risk perception and work behaviour in forestry: implications for accident prevention policy. Accid. Anal. Prev., 189—200.

ABSTRACTS

Ten Years Experience of Accident Registration

ARNE RASMUSSEN

The Danish Labour Inspection Service, Statistical Department, Postbox 858, DK 2100 Copenhagen (Denmark)

In 1958 registration of accidents was begun by The Labour Inspection Service in Denmark. In 1973 the system was changed to a system the Service is still using. It has two aims: (1) Prevention of accidents, and (2) Providing statistics. The registration system contains information of the place of employment, the work process, the accident and the injured person. These 4 main parameters are further divided into details.

An example: In a shipyard an accident happened in a workshop. The injured person served a drilling machine. He tripped over a piece of iron while he was cleaning the machine. He broke his leg.

Industry = Shipyard
Type of work = Machine work
Kind of work = Cleaning of machines
Technical factor = Drilling machine
Technical detail = Loose object
Technical event = No technical event
Type of accident = Trip over an object
Part of body = Leg
Kind of injury = Fracture.

The output selection technical factor (= drilling machine) shows accidents by work processes, where drilling machines are used, but not *only* accidents *caused by drilling machines*. If we want output of accidents attributed to drilling machines, it is necessary to select by technical detail "all parts of drilling machines", too, because the part contributed to the accident if it caused a disturbance or deviation which caused the accident.

It is not possible to register the accident in both ways, but it is possible to obtain an output aimed at the two goals.

How to Use the Information System on Occupational Injuries (ISA) in Research

JAN CARLSSON

National Board of Occupational Safety and Health, S-171 84 Solna (Sweden)

ISA contains information about all occupational accidents in Sweden with at least one day of absence and information about reported occupational diseases. ISA's aim is to supply information for prevention of occupational injuries.

The information is classified and registered in a database. All occupational forms are also registered on microfilm. The system is useful in research for:

- quantification and identification of occupational injury problems using data for persons, occupations, workplaces, accidents, situations, etc.
- analyses or consequences of injuries such as days of absence, type of injuries.
- analyses on connections between, e.g., environmental factors and type of injuries.
- epidemiological studies through, e.g., comparison between ISA data and demographic data.

Occupational Accident Data and Safety Research in Finland

PEKKA MAIJALA

Technical Research Centre of Finland, Occupational Safety Engineering Laboratory, Box 656, SF-33101 Tampere 10 (Finland)

The information on occupational accidents available in Finland generally tends to describe and categorize the results of accidents rather than the events which preceded them.

It is not currently possible to isolate the major causes (that is, contributing events and conditions) of these accidents, or to introduce possible preventive measures on the basis of the available statistics.

The most important information systems of occupational accidents in Finland are: official statistics of Finland, statistics of insurance companies, a register of serious occupational accidents, special statistics of type of business, statistics on company level.

All the information systems have their own basic data which have been gathered by using different methods. When we use these information systems in our research projects we have to know the reliability of the information we collect and how we can combine these systems to improve their usefulness.

The Analysis of Injuries Amongst Workers in a Hospital for Chronic Patients

MONIQUE LORTIE

Ecole Polytechnique de Montréal, Génie Industriel, C P 6079 Succ A Montréal, P Q, H3C 3A7 (Canada)

532 injuries were compiled from the 225 records of nursing aids working in a hospital for prolonged health care. Injuries were classified in two categories.

An analysis matrix was developed for each category and the data was codified and treated by computer. In parallel with this phase, an analysis of the nursing aids' work was made.

Sexual differentiation of the injuries' profile was demonstrated with the results: men and women do not experience the same types of injuries and the localization of the injuries differs.

Analysis shows that, for the same work task, men accomplish their duty differently from women. Relations between these two types of results is discussed. A hypothesis regarding the types of efforts and postures which could lead to a larger risk of injuries is stated.

Injury Information Systems for Management

D.B.L. VINER

18 Malmsbury St., Hawthorn, Victoria 3122 (Australia)

A means of classifying historical injury data is described which offers clear definition of types of occurrence leading to the injury. The approach is based on the energy-damage model and the risk estimation model of Rowe (1977). (An Anatomy of Risk. Wiley Interscience, New York.)

The developed concept enables injurious occurrences to be classified with a high degree of repeatability — a feature not possible with the ILO accident type classifications. The first two levels of classification contain categories of international applicability to all industry types. The third level is industry-specific.

Principles of analysis of this data are briefly discussed.

RESEARCH STRATEGIES AND METHODS APPLIED TO
FALL ACCIDENTS

ACCIDENTAL FALLS AT WORK, IN THE HOME AND DURING LEISURE ACTIVITIES

JOHAN LUND

State Institute of Consumer Research, Box 8004, Dep., Oslo 1 (Norway)

ABSTRACT

Lund, J., 1984. Accidental falls at work, in the home and during leisure activities. *Journal of Occupational Accidents*, 6: 181—193.

A study of accidental falls was carried out, especially with regard to falls in the home and during leisure time. Other falls were reviewed through literature studies. The study is based on about 4500 falls on the same level (of which about 1500 were slipping accidents) recorded in hospitals in 4 Nordic countries during the years 1977—1980, when collecting details of home and leisure-time accidents.

The total number per year and the distribution of accidents are anticipated for the Nordic countries from accidents which were treated in hospitals (in- and out-patients). Falls on the same level and slipping accidents are analysed for their main characteristics and a number of preventive measures are suggested.

INTRODUCTION

This paper is a resumé of part of a Nordic project about accidental falls, particularly falls on the same level (Lund, 1983). The project is based on a thorough study of the literature on falls on the same level, especially including slipping accidents. The literature was mainly based on accident victims registered in hospitals as in- and out-patients, but investigations based on other sources were also studied, such as industrial accident statistics and accidents in institutions. In addition, literature concerning fall mechanisms, characteristics of shoes and floors was studied. In the report on the project there is also an analysis of approx. 4500 falls on the same level in homes and during leisure-time activities in the Nordic countries. The project was initiated by the Nordic Council's Executive Committee for Consumer Affairs.

The Consumer and Product Control Authorities in the Nordic Countries are working towards establishing an accident registration system in hospitals. In 1977 a method study was carried out, collecting data concerning approx. 30,000 accidents from 20 hospitals in all Nordic countries (Nordisk Ministerråd B 1978). The biggest group of accidents according to the E-code (External Cause of injury) was due to falls on the same level (E-885). The size of this group varied from 18 to 36% of the home and leisure-time accidents in the five different Nordic studies.

The E-code is an international injury classification (WHO 1967).

The objective of the Nordic project (Lund, 1983) was to gain knowledge about the causes and course of falls on the same level in order to be in a better position to prevent them. An additional intention was to provide knowledge which could be useful in the Nordic work on developing methods of measuring slipping safety.

It is important to realize that in this project, and others referred to in this paper, accidents are defined as an event causing people to seek medical treatment for injuries or what they or their parents consider to be an injury. All degrees of severity, from death to no injury at all, are involved. One of the criteria in our definition of "accident" is therefore the seeking of medical treatment. It is also important to emphasize that medical treatment received at work is *not* included in the registration of these accidents.

CHARTING THE EXTENT OF ACCIDENTAL FALLS

In another Nordic project (Lund, 1983a) it was calculated that every year about 13% of the population in the Nordic countries seek medical treatment for injuries sustained in an accident. Table 1a lists the studies which have led to that conclusion. Accidents requiring medical treatment at work or at school are not included. The table also gives the size of the various accident groups in these studies. Special accidents are violence, and injuries inflicted by animals, insect bites, etc.

TABLE 1A

Incidence of accidents and size of accident-groups found in different studies in the Nordic countries. (Incidence is the proportion of the population which each year seeks medical treatment due to an accident. N is the number of patients with accidents in the different registration projects)

Studies	Calculated incidence (%)	N	Accident groups (%)			
			Occupational	Transport	Special	Home, leisure
Århus, Randers, DK[a] 1 year 1982	n/a	48,055	13	14[g]		72
Mikkeli (SF)[b] 1 year 1980/81	15	8,894	29	6	5	59
Oslo (N)[c] 1 week 1973	16	1,425	20	5		76
Rogaland (N)[d] 4½ months 1977	14	7,916	24	6	7	64
Falköping (S)[e] 1 year 1978	11	3,641	20	8		73
Østgötaland (S)[f] 4 weeks 1982	11	2,264	17	4		79

TABLE 1B

Calculated incidence and size of accident groups in the Nordic countries

Denmark (DK)	14	15	11	7	67
Finland (SF)	14	30	5	5	60
Iceland (IS)	14	30	5	7	58
Norway (N)	14	22	6	7	65
Sweden (S)	11	18	6	8	68

[a] Information from B. Sieberg, Århus hospital.
[b] Honkanen et al., 1983.
[c] Reigstad, 1976.
[d] Lund, 1982.
[e] Schelp et al., 1979.
[f] Konsumentverket 1982.
[g] Bicycle accidents are included here. In other studies they are in the home, leisure group.

In Table 1b, we show the anticipated incidence in the Nordic Countries, together with an anticipated distribution over the different accident groups.

In Table 2, we have calculated the anticipated number of people in the different Nordic countries and the total number of people who will be treated in the health system each year for injuries sustained in an accident. It is quite a considerable number of accidents.

We have endeavoured to find the share of falls in the different groups by a study of the literature. There are very few studies which give a survey of all kinds of accidents, the share of falls in general and slipping accidents in particular, in one district.

TABLE 2

Anticipated number of persons in the Nordic countries receiving medical treatment due to injury sustained in an accident. The base for the anticipation is the calculated incidences and sizes of the different accident groups from Table 1b

Nordic countries	DK	SF	IS	N	S	Total	
Calculated population (1,000)	5,104	4,752	235	4,059	8,278	22,428	
Assumed number accidents total	715,000	666,000	33,000	570,000	910,000	2,894,000	(100%)
Home and leisure	480,000	400,000	19,000	370,000	619,000	1,888,000	(65%)
Occupational	107,000	200,000	10,000	125,000	164,000	606,000	(21%)
Transport	79,000	33,000	1,500	34,000	55,000	202,500	(7%)
Special	49,000	33,000	2,500	41,000	72,000	197,500	(7%)

Tables 3, 4 and 5 show the size of falls on the same level in all accidents, in home and leisure accidents and in occupational accidents in some investigations mainly from the Nordic countries. We notice that in general, the variations are quite small from one study to another.

TABLE 3

All accidents, amount of fall accidents and falls on same level from different registrations/investigations of people treated in hospitals

Place, country	Registration period (months)	Number of accidents		
		Total	Falls	Falls on same level
Tölö, SF[a]	1979 (12)	28,460	10,744 (38%)	6,088 (21%)
Stavanger, Sandnes[b] N	1978/79 (12)	5,059	1,839 (36%)	n/a
Odense, DK[c]	1980 (12)	31,948	10,385 (33%)	6,091 (19%)
Umeå, S[d]	1978/79 (12)	ca. 8,000	2,650[d] (33%)	1,718 (21%)

[a] Hospital records, unpublished
[b] Lund, 1984
[c] Statistical report (preliminary) Accident Analysis Group, Odense University Hospital
[d] Ulf Bjørnstig unpublished material. Falls during skiing and skating are *not* included, therefore falls are underrepresented.

TABLE 4

Home and leisure accidents, number of accidental falls and falls on same level from some registrations/studies of people treated in hospital

Place, country	Registration period (months)		Number of accidents		
			Total	Falls	Falls on same level
Odense, Randers Århus[ac] DK	1/4—31/10 1977	(7)	11,496	3,834 (33%)	2,089 (18%)
Helsinki, Pargas Iisalmi, Lahti[a], SF	17/1—30/9 1977	(8½)	13,974	7,491 (57%)	4,709 (36%)
Stavanger, Sandnes[b], N	1/6/77—31/5/78	(12)	5,231	2,743 (52%)	1,477 (28%)
Umeå, Uppsala[a], S	1/1—30/9/77	(9)	4,358	1,851 (43%)	997 (23%)

[a] Nordisk Ministerråd 1978
[b] Lund 1981
[c] Accidents among the aged, and therefore the falls, are underrepresented.

Based on these tables, we have made one figure illustrating the extent of falls in the main accident groups (Fig. 1).

From Table 3, we assume that falls on the same level account for 21% of all these accidents. The rest — 16% — is constituted by falls between levels, on stairs, etc. Slipping and tripping will occur in both groups. In the material from Tölö-hospital (6088 falls on the same level — see Table 3) 69% were slips and 31% were tripping accidents. In the material from Stavanger/Sandnes, we found 8% slips and 14% tripping in the other accidental falls —

TABLE 5

Occupational accidents, number of accidental falls and falls on same level from different registrations/studies

Place, country	Registration period (months)	Number of accidents		
		Total	Falls (%)	Falls on same level
Tölö[a] SF	1979 (12)	4,581	24	554 (12%)
Malmö[b] S	1974 (2)	324	17	25 (8%)
Sweden[c]	1980 (12)	116,104	15	12,435 (10%)
Århus[d] DK	1979/80 (12)	4,932	13	350 (7%)
Oslo[e] N	1973 (3)	2,470	21	271 (11%)
Germany[f]	1975 (12)	1,545,000	15	n/a

[a] Hospital records (unpublished)
[b] Anderson et al., 1975 (hospital records)
[c] Sveriges officiella statistikk, Stockholm 1983[x]
[d] Hospital material, unpublished
[e] Reigstad 1978 (hospital material)
[f] Abt 1976 — German official occupational accident statistics[x]

[x] This registration is based on reports to the labour accidents insurance system, which is a different definition of an accident from the hospital-based registrations.

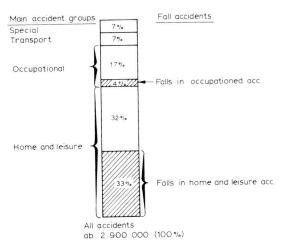

Fig. 1. Proportions of accidents treated in the Nordic health system (except health systems in schools and at work) divided into main accidents groups (from Table 2), all types of falls in the occupational group (approx. 20% of all occupational accidents, Table 5) and in home and leisure group (approx. 50% of all home and leisure accidents, Table 4).

E code 880-884 and 886-887 (Lund, 1983 pages 34 and 35). Figure 2 is based on those results. It must be emphasized, however, that Fig. 2 is based on two registration projects in hospitals, and it might be difficult to distinguish if slipping or tripping occurred in the fall. Further investigations, and especially in-depth investigations, are necessary to verify Fig. 2.

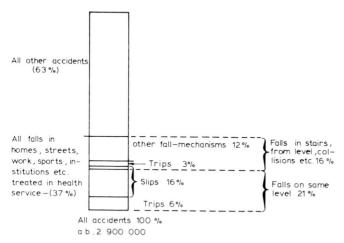

Fig. 2. Proportion of different fall accidents compared with all accidents treated in the Nordic health system (except health system in schools and at work). The size of the different fall accidents is based on a Finnish and a Norwegian investigation, and is approximate. The Figure is more an illustration than an exact reproduction of the Nordic situation.

Due to the number of accidents from these two figures it is, however, obvious that falls constitute a type of accident that is a most considerable burden on the public health system and which results in personal pain for many people in the Nordic countries.

A Danish study (Søgaard, 1981) calculated the cost of different accidents to society in 1979 (Table 6). The costs are: lost production due to deaths,

TABLE 6

Costs of accidents as calculated by Søgaard (1981)

	Million Danish kroner	Proportion (%)
Falls	1,083	32
Traffic accidents	1,080	32
Other transport accidents	179	5
Poisonings	194	6
Other accidents	816	24
	3,352	99
Costs impossible to allocate	921	
Total cost of accidents	4,273	

disablement, time for treatment and sick-leave, and costs for treatment in the health system.

These costs can to some extent also be considered as a severity index, as they reflect length of treatment and loss of production due to deaths and disablement.

This calculation shows that the cost for falls are of the same size as for traffic accidents. This is mainly due to the long treatment for old people with hip fractures. However, we know that falls are more numerous than traffic accidents, which means that, on average, each traffic accident is more costly than each fall.

ANALYSIS OF APPROX. 4,500 FALLS ON THE SAME LEVEL IN (HOME AND LEISURE-TIME)

In analysing the accidents, we used an accident model as shown in Fig. 3. We used four main groupings:

- biographical data
- characteristic data from the moment immediately before the accident
- characteristic data from the circumstances of the accident itself, the chain of events
- characteristic data from the circumstances of injury.

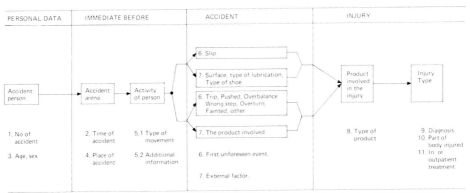

Fig. 3. Accident model for fall-accidents on the same level with variables.

We analysed two different sets of data (Table 7 shows the data analysed):

- falls on the same level
- slipping accidents as a proportion of the falls on the same level.

Because our accident descriptions were relatively scarce with regard to information concerning the first unforeseen event, we did not get so much information as we had hoped from the analysis. In-depth investigation, with on-the-spot examination just after the accident, is necessary.

TABLE 7

Data analysed in the Nordic project on falls on same level (Lund, 1983)

Place where data collected	Registration period (months)	Number of accidents		Type of accidents
		Falls on same level	Slipping	
Århus[a]	February 1980[a]	429	149 (35%)	Home and leisure
Helsinki, Pargas Iisalmi, Lahti[b]	17/1—30/9/77 (8½)	895	299 (33%)	Home and leisure
Stavanger, Sandnes[c]	1/6/77—31/5/78 (12)	1,485	246 (23%)	Home and leisure
Umeå[d]	1/11/78—31/10/79 (12)	1,718	822 (48%)	All accidents except skiing and skating accidents

[a] Data collected especially for this project as a pilot study.
[b] Every 5th fall accident was taken from data collected earlier (see Table 4). On checking, 46 of the 941 accidents were not falls on same level, which leaves 895.
[c] Data collected earlier (see Table 4). On checking we found 8 more falls on same level, which give a total of 1485. Under-representation of accidents in months October—January. Hospitalized patients over-represented. Area has a mild climate.
[d] Data the same as referred to in Table 3.

Analysis of accidental falls on the same level

Examination of the data on falls on the same level, particularly the Swedish, Norwegian and Finnish data, showed that falls in these three places have a number of common features.

Table 8 shows the most important findings from the data from the three countries. When studying this table, however, the limitations of the data, given in footnotes, must be emphasized. The data from the three countries are not directly comparable, due to differences in the sample background. Being aware of those limitations, we can however assume certain characteristics for falls occurring on the level in the three Nordic countries, Norway, Sweden and Finland:

- Size of age group. For children and old people, this is almost the same as for young people and adults (Table 8 — point 1).
- The first unforeseen event. Children are prone to falls particularly through stumbling and slipping during play. Young people fall mainly through loss of balance and stumbling during sports, athletics and play. Adults fall through slipping and loss of balance during walking, sports and athletics. The elderly are particularly prone to slipping accidents.
- The main places for accidental falls on the same level are at home and on the street, road, etc.
- Most of the people involved in falls walk normally immediately before the accident occurs. It is particularly these individuals who slip. The

others have a more varied accident pattern through stumbling, taking a false step, losing balance, etc. Naturally, it is mainly the surface underneath them on which people are hurt in the fall. They either hit the ground if they are outside, or the floor and/or furniture when indoors.

It is evident that these injuries can be quite serious: there are many fractures and sprains among them. Particularly among the elderly, this also results in hospitalization.

TABLE 8

Falls on same level. Main characteristics of data collected in three different countries.

		Percentage of total number of accidents in the data		
		Norway (%)	Sweden (%)	Finland (%)
Distribution of age groups:[a]				
Children	0—15 years	50	33	34
Young people and adults	16—64 years	31	48	50
Old people	65 years +	19	19	16
Total		100	100	100
The first accident in each age group:[b]				
Children 0—15 years	— Stumbling	36	33	66
	— Slipping	18	34	15
Young people 16—24 years	— Loss of balance	27	35	6
	— Stumbling	24	18	59
Adults 25—64 years	— Slipping	37	58	37
	— Loss of balance	25	24	2
Old people 65 + years	— Slipping	47	54	50
Place where falls occurred:				
Homes, inside building		25	25	17
street, road		23	27	34
Movement just before the fall				
Walking normally		43	55	55
Proportion who slipped		68	63	49
Injury[c]				
Fractures		48	33	29
Sprains		18	28	31
Total number of accidents in the data		1485	1718	895

[a] Norwegian data are home and leisure accidents, i.e. the adult group (industrial) is under-represented. The Swedish data have industrial accidents included, but ski- and skating accidents are excluded, i.e. the adult group is over-represented. The Finnish data under-represents children.
[b] The differences between the size of the group may also be caused by different code practices.
[c] The Norwegian data over-represents hospitalized persons, which may explain the high ratio of fractures.

ANALYSIS OF SLIPPING ACCIDENTS

Table 9 gives the main characteristics of slipping accidents. They are mainly a problem for adults and elderly people. Women are over-represented in this group in contrast to other types of accidents, where the opposite is the case.

TABLE 9

Slipping accidents. Main characterstics of data collected in three different countries. — For footnotes, see Table 8.

	Percentages of total number of accidents in the data		
	Norway (%)	Sweden (%)	Finland (%)
Distribution of age groups:[a]			
Children 0—15 years	22	11	7
Young people and adults 16—64 years	49	69	68
Old people 65 + years	27	19	24
Total	100	99	99
Sex distribution[a]			
Women	61	52	58
Men	39	48	42
Total	100	100	100
Where accident happened			
Inside	27	25	16
Outside	73	75	84
Total	100	100	100
Main type of surface			
Snow — Ice	73	57	74
Usual movement before a slip			
Walking normally	68	70	81
Injuries[c]			
Fractures	65	40	42
Sprains	13	22	22
Total number of accidents in the material	346	822	299

Slipping accidents most commonly take place outdoors in the Nordic countries, and they are strongly connected with snow and ice. The main movement before a slip is walking normally.

Slipping accidents often cause fractures and sprains. The hospitalization frequency seems to be the same as for other falling accidents on the same level, i.e. somewhat higher than for other home and leisure-time accidents.

The representativeness of the results

The results from these three groups of data show quite a lot of common characteristics. The data are collected from a western, mild part of Norway, a northern part of Sweden and a southern part of Finland. It is difficult to judge if this data is representative of those three Nordic countries. The limitations of the data must also be considered (footnotes in Table 8). However, even with these limitations, we can see some broad lines of similarities in Tables 8 and 9. A rough mean of most of the values could be a Nordic mean of the characteristics of those accidents. We need, however, more thorough and complete studies of these types of accidents from more places in the Nordic countries to be more sure of this assumption.

In the next section, we will propose a number of preventive measures which have come to mind while studying these data. It is not possible to relate the proposals to the number of accidents. We do not know how many falls on snow and ice would have been prevented with better snow clearing and gritting. This would require intensive in-depth studies. However, if the proposed preventive measures are carried out, it is obvious that a lot of falls will be avoided.

SUGGESTIONS FOR PREVENTIVE MEASURES (HOME AND LEISURE ACCIDENTS)

These suggestions are directed towards the relevant authorities, producers and consumers.

Road/Highway Authorities. The most important preventive measure for slipping accidents is probably an efficient system of snowclearing and gritting in towns and rural districts. The aspect of clearing pavements, roadsides, etc., should be considered in the planning of road construction.

Building authorities. Critical sites for pedestrians should be equipped with solidly built handrails. Melting water from roofs should be led into the gutter in such a way that patches of ice are avoided. National building codes should include measures for the prevention of fall- and slipping accidents.

Social authorities. Elderly people are especially prone to falls. Institutions for the elderly and for convalescents should consider this problem. Instruction should be provided for the staff in such institutions and for people helping the elderly in their homes.

Shoe manufacturers. Types of shoes suited for slippery outside surfaces should be developed. Shoes for elderly people are a special challenge. Many indoor falls are probably due to indoor shoes which are badly fitted and have slippery soles.

Developers of accessories for shoes. There is a need for the development of an aid to be used on slippery surfaces. It should be noted that Nordic industrial designers are interested in co-operation with the relevant authorities in a development project along these lines.

With regard to shoes, there is a need for better product information and distribution of such material. The producers of shoes should provide information concerning sole material and the relevant surfaces. Shoe retailers should have more information about the shoes they are selling.

The efforts of the Nordic consumer authorities with regard to requiring standardized, specified labelling should be intensified.

The floor cleaning industry. One should investigate how this industry is connected with the prevention of slipping accidents.

Producers of bathroom floors. Falls in bathrooms are a special problem.

Indoor floor covering industry. One should discuss with the industry the prevention of falls through development of safer coverings.

Landlords/owners of apartment houses. They should be informed of the rules concerning the clearing of snow from, and gritting of, the pavement.

Elderly people. They should be informed of how accidents happen, given advice on shoes, aids, preventive measures in the home, and physiological conditions which change with age.

The general public. Consumers should be informed of the frequency and pattern of accidental falls in general, shoes, clearing of pavements, cleaning of floors, polishes, paint and varnishes, falls among children.

The following research and development projects should be initiated:

- A registration system for accidents in the home and during leisure-time.
- In-depth investigations concerning: (1) outdoor accidents on snow and ice and (2) indoor accidents.
- Testing methods should be evaluated critically. One should try especially to find a method for measuring slipping safety on ice and snow.
- Problems of elderly people. Research for elderly people, concerning nutrition, physical activity and other aspects relevant to their age and accident-proneness should be carried out.

A four-stage model for the spreading of knowledge about falls on a Nordic/national basis is suggested:

- Distribution of reports/synopses to involved authorities, institutions, industries.
- A seminar for specialists/authorities.
- A seminar for key persons/contact persons for consumers.
- Information campaign.

The creation of local action groups in municipalities which can work towards preventing falls in the local environment is suggested.

In addition, resources should be set aside by the Nordic Council's Executive Committee for Consumers Affairs to ensure that initiative will be taken to follow up some of these suggestions. One could envisage a Nordic action group being established for this purpose.

REFERENCES

Abt, W., 1976. Stolper-, Rutsch- und Sturzunfälle in den Jahren 1974 und 1975. Die Berufsgenossenschaft 11: 429—435.

Anderson, R., Johansson, B., Lindén, K., Svanström, K. and Svanström, L., 1975. Om olycksfall i arbetet. (On occupational accidents — in Swedish). Malmø, 208 pp.

Honkanen, R., Korhonen, A., Korhonen, K., Koivumaa, H., Pursiainen, M. and Jolkkonen, I., 1983. Lääkärissäkäyntiin johtaneet tapaturmat Mikkelin seudulla 1980—81. (Medically attended accidents in Mikkeli 1980—81 — in Finnish with English summary.). Kansaneläkelaitoksen Julkaisuja ML: 32 1982. Helsinki, 145 pp.

Konsumentverket 1982: Öka säkerheten I. Pilotprojekt (Increase Safety I. Pilot project — in Swedish.), 34 pp.

Lund, J., 1981. Ulykker i hjem og fritid. Metoder for registrering og analyse av 5231 ulykker innsamlet i løpet av ett år (1977/78 - Rogaland II). (Accidents in home and leisure. Methods for registration and analysis of 5231 accidents during one year 1977/78, in Norwegian with English summary.) Statens institutt for forbruksforskning. Report no. 59. 120 pp.

Lund, J., 1982. Accidents in one region of Norway during a period of one year. J. Occupational Accidents 4: 245—256.

Lund, J., 1983. Fallulykker på samme nivå i hjem og fritid i Norden. (Fall accidents on same level in home and leisure in the Nordic countries. In Norwegian with English summary.) Nordisk Embetsmannskomite for konsumentspørsmål (NEK). Report no. 4, Oslo, 240 pp.

Lund, J., 1983a. Registrering av ulykker. Oppfølging av et tidligere nordisk prosjekt (Registration of accidents. Follow up of an earlier Nordic project — in Norwegian with English summary). Nordisk Embetsmannskomite for konsumentspørsmål (NEK). Report no. 16, Oslo, 165 pp.

Lund, J., 1984. Ulykker i hjem og fritid. Registreringssystem for ulykker som behandles ved sykehus/primærhelsevesenet — Rogaland IV (i trykken) (Accidents in home and leisure time. Registration-system for accidents treated at hospital/primary health system — in Norwegian.) Statens Institutt for forbruksforskning, Oslo. In print.

Manning, D.P., 1974. An accident model. Occupational Safety and Health (4): 14—16.

Manning, D.P. and Shannon, H.S., 1979. Injuries to lumbosacral region in a gearbox factory. J. Soc. Occup. Med. 29: 144—148.

Nordisk Ministerråd 1978. Inrapportering av olycksfall i hemmen och deras grannskap. (Registration of accidents in the homes and their neighborhood) Del II A Danmark (in Danish) 16 pp., Del II B Finland, 24 pp., (in Swedish), Del II E Sverige (in Swedish). Stockholm, 33 pp.

Nordisk Ministerråd B, 1978. Innrapportering av olycksfall i hemmen och deras grannskap. (Registration of accidents in the homes and their neighborhood — in Swedish with English summary). Report no. 14. Stockholm, 57 pp.

Reigstad, A., 1978. Ulykker i arbeidsmiljøet. (Accidents in occupational milieu — in Norwegian, English summary). Oslo, 182 pp.

Schelp, L., Svanström, K. and Svanström, L., 1979. Olycksfall i Falköpings kommun, år 1978. (Accidents in Falköping municipality — in Swedish) Landstingets Hälsovård, Skövde.

Sveriges officiella Statistik, 1983. Arbetsskador 1980. (Official statistics of Sweden: Occupational Injuries 1980 in Swedish with English summary). Stockholm, 180 pp.

Søgaard, J., 1981. Samfundsmæssige omkostninger ved ulykker (The cost for society due to accidents — in Danish) i "Ulykker — Forebyggelsesrådets konferanse om forebyggelse av ulykker". Indenriksministeriet, København, pp. 99—125.

WHO (World Health Organisation), 1967. International Classification of Diseases, 8th revision. Official records 160, 9, 11, Annex 18 (E-code), Copenhagen.

ABSTRACTS

Advantages and Limitations of Various Methods Used to Study Occupational Fall Accident Patterns

H. HARVEY COHEN

Safety Sciences, Division of Syncor International, 7586 Trade Street, San Diego, CA 92121 (U.S.A.)

This paper presents a discussion of over 6 years of research performed by the author and funded principally by the National Institute for Occupational Safety and Health (NIOSH) in the U.S. Three major studies have thus far looked at falls on level working surfaces, falls on stairs, and falls from ladders. To a lesser extent, in other studies, falls from other elevated workstations have been looked at. A combination of case history, epidemiologically-based, and otherwise experimentally controlled, retrospective and prospective field study approaches have been used, including: 1) Review of existing injury records, 2) Detailed accident investigations, involving both interviews and site observations, 3) Comprehensive, case-control, retrospective interview and site observation surveys, 4) Prospective longitudinal studies involving both setting up and operating for several years safety and health information systems for specific high-risk industries, and 5) Video recorded observations of critical incidents as they actually occurred over extended periods at high risk work sites. Findings to date are discussed in light of the types of data best obtainable from each approach.

The Accident Model Applied to Back Injuries

J.D.G. TROUP

University of Liverpool, Department of Orthopaedic and Accident Surgery, Royal Liverpool Hospital, P.O. Box 147, Liverpool I69 3BX (United Kingdom)

401 cases of back pain occurring during 1980 were investigated. True accidents were separated from non-accidental injuries (NAI), and back pain of unknown cause. Accident information was recorded on diagrams of the accident model. Structuring of information in this way facilitated the study of first unforeseen events, body movements and other factors contributing to the accidents.

80 (66%) of the first unforeseen events were underfoot, including 57 slips. There were highly significant differences between the body movements contributing to accidents and NAI, and significantly more of the NAI than accidents involved handling of loads. 52% of the sample of accidents and NAI were not handling loads.

Accident Analysis, Biomechanics, and Tribology for Slipping and Falling Injury Prevention

LENNART STRANDBERG

Accident Research Section, National Board of Occupational Safety and Health, S-171 84 Solna (Sweden)

With the new Swedish "sequential" accident registration model slipping appears in more than 11% of the occupational accidents. Since most slipping accidents occur on contaminated surfaces, only two of 13 shoe(—lubricant)— flooring combinations were selected — dry and clean — in an interlaboratory comparison of some slip-resistance testing apparatus. The friction readings from nine meters were compared with friction values computed from walking experiments. The correlation coefficients between the apparatus and the walking experiments varied substantially (from −0.32 to +0.98). The greatest correlations were achieved with testing methods, simulating closely the force and motion time histories in human gait according to late biomechanical measurement data.

Some Aspects on Mechanical Testing of Fall Safety Devices

H. ANDERSSON

Division of Solid Mechanics, National Testing Institute, Box 587, S-501 15 Borås (Sweden)

In order to modernize standardized test methods as effectively as possible it is important to prepare technical arguments which make use of the most recent developments.

Modern measurement techniques are used together with modelling of actual biomechanical properties of tissue to evaluate the interaction between impact and shock damping properties — on the one hand in the system safety helmet/brain, and on the other hand in the system of a fall safety belt and the whole human body. In the case of safety helmets it appears that rotational movements are an important source of injury.

Research in Fall Protection at Ontario Hydro

A.C. SULOWSKI

Ontario Hydro Research Div., 800 Kipling Avenue, Toronto, Ontario, M7 5S4 (Canada)

Falls from heights are the second largest contributor, after electrical contacts, to accidents at electric power utilities. In 1977, the Research Division of Ontario Hydro embarked on comprehensive R&D work in fall protection in order to permit the Corporation to upgrade the existing level of protection against falls. The research work included the development of new test methods, establishment of a rigid weight to manikin conversion factor and both a theoretical analysis and application-oriented system design. The theoretical analysis resulted in the development of a new formula for maximum arrest force and a simplified graphical method applicable in the design of fall arresting systems. Various available equipment was evaluated and a new device was designed and invented. Several special purpose systems were designed, tested (laboratory and field) and introduced in operational divisions, including systems for transmission lines, construction and maintenance, station maintenance, boiler maintenance, wood pole climbing (underway), and many others. Dissemination of the information generated in the R&D work included seminars, films, video programs, research reports, papers, etc. The paper contains the highlights from each of the abovementioned types of work.

Accidents Involving Falls from Roofs — Survey and Technical Preventive Measures

PER-OLOF AXELSSON

Occupational Accident Research Unit, Royal Institute of Technology, S100 44 Stockholm (Sweden)

Accidents involving falls are mong the commonest types of occupational accident. Falls to a lower level frequently result in serious injury and occasionally death. Many of the accidents involving falls to a lower level take place while working with or on the roofs of buildings.

The results of a survey of problems relating to work on roofs, involving statistical processing and observations made while conducting studies at worksites, are described. The survey has formed the basis for technical development work which has led to proposals for solutions of permanent barrier holders which are installed in new buildings and left in place for future maintenance and reconstruction work.

SAFETY EFFECTS OF NEW PRODUCTION TECHNOLOGY

A STATISTICAL STUDY OF CONTROL SYSTEMS AND ACCIDENTS AT WORK

TOMAS BACKSTRÖM and LARS HARMS-RINGDAHL

Occupational Accident Research Unit, Royal Institute of Technology, S-100 44 Stockholm (Sweden)

ABSTRACT

Backström, T. and Harms-Ringdahl, L., 1984. A statistical study of control systems and accidents at work. *Journal of Occupational Accidents*, 6: 201—210.

One purpose of the study was to get an estimate of the number of occupational accidents in Sweden in which a control system was involved. Another purpose was to get a picture of the structure of these accidents. The main source of information was the Swedish data system of occupational injuries.

The number of accidents in connection with control systems has been estimated to be around 2000 per year, corresponding to 2% of the occupational accidents in Sweden, but for some types of equipment accident rates are higher.

177 accidents were selected and studied. The most frequent types of equipment involved were presses, transportation systems and cutting machines. Inadvertent start caused 84% of the accidents.

In 19% of the accidents technical faults were observed. This figure was higher for some types of equipment e.g. transportation systems (28%). In one third of the accidents the operator had corrected a deviation in the material or the equipment. Thus, hazards in this kind of task seem to be a major problem. In 55% of the accidents the operator caused the inadvertent start.

Some recommendations are given on the design of automatic systems.

INTRODUCTION

Increased automation in industry means that control systems are taking over a larger part of the direct control of equipment. Some dangerous tasks disappear and new ones are created. Experience from accident research in, for instance, the paper industry (Harms-Ringdahl, 1982 and 1983) showed many defects in safety aspects of control systems. Another impression is that inadequate attention is paid to safety aspects of the control systems at the design stage. Proposed safety measures were met by arguments such as that they were not necessary since there was no reason for people to be in the danger zone. There is also a pessimistic view on interlocks — they will be put out of action.

This experience has led to an interest in developing the safety philosophy in the design of control systems. Therefore we were interested in surveying accidents in connection with control systems. How frequent are such accidents? What happens? Are inoperative interlocks a common cause of accidents? Studies of inadvertent starts have been made earlier (Faxö, 1981 and Rouhiainen, 1982), but they only partially answer these questions.

In this study a control system is defined as a device which automatically controls different active states, e.g. machine movements.

The selection criterion for the accidents has been that a control system should have controlled the energy causing the injury, e.g. a machine starting inadvertently or not stopping as intended. This means that inadvertent starts with no control system involved, e.g. with a line switch, have been excluded.

COLLECTION OF INFORMATION

The main source of information has been the Swedish Occupational Injury Information System (Andersson and Lagerlöf, 1983). However, there was no direct way to obtain answers to the questions. The starting point was to select accidents by means of four codes. In a simplified translation, the codes were:

- unintentional start via control button
- other unintentional movement of machine
- functional disorder
- using wrong control button

From these, the most severe accidents were studied. Cases involving a control system were selected.

Another way of finding relevant accidents was to use codes for control systems. However, these only made a small contribution. Special codes for industrial robots and multi-operation machines were also used. A number of cases also came from a manual scan of randomly chosen reports. A total of 177 accidents involving a control system in 1981 was collected.

A general deficiency in this material is that the descriptions of the accidents are short, and that information can be missing. Another weakness is that there are systematic errors due to the methods of selection. Distribution according to the equipment involved is probably deceptive; for example, accidents with chemicals or hot material will not be detected.

44 accidents investigated by the Labour Inspectorate in 1980 were also selected. These cases have been separately treated and are not incorporated in the tables.

Among the studied accidents found in this way, there were 3 fatalities. Of 46 fatal accidents investigated by the Labour Inspectorate, 4 were connected with control systems.

NUMBER OF ACCIDENTS

An estimate has been made of the number of accidents per year in which a control system was involved. The starting point was the number of accidents (N) coded as inadvertent start, etc. In order to get the number (C) involving control systems, corrections must be made. For that purpose the following expression was used:

$$C = N \times k_a/k_b$$

k_a corrects for the fact that only a fraction of the selected accidents are connected with a control system. The factor was found by studying a sample of 413 accidents. 124 of these were relevant, giving k_a a value of 0.30. k_b corrects for the fact that not all accidents with control systems will be found by the codes used. The factor was estimated by checking accident reports found in a different way. The largest part came from a manual scan of randomly selected reports. Of 44 accidents, 12 had one of the actual codes. This gives k_b the value 0.27.

The obtained values thus give an estimate of 1,800 accidents per year with control systems, with a standard deviation of 30%. However, care must be taken in the interpretation of this figure because of the large corrections, the large standard deviation, and the systematic errors.

ANATOMY OF THE ACCIDENTS

177 accidents were studied and classified in different respects. Table 1 shows the types of event that led to the injury. Except for two cases, the injuries were caused by a moving part causing contusion, cuts, etc.

TABLE 1

Type of event causing accident (sample size 177)

Event	Frequency (%)
Inadvertent start	83
Failure to stop	5
Intended movement	11
Abnormal movement	1
Total	100

Technical deviations in the control system and deviations in the operation have been studied. The terms are further developed below. A summary of the deviations is shown in Table 2.

In 55% of the accidents, an operational deviation existed. For technical deviations the figure is 32%. In the group "No deviation" the control system worked normally and the operator did not make a direct error in connection with the control system. These cases were often detected through proposed countermeasures in connection with control systems. They should be included in the sample according to the definition.

TABLE 2

Occurrence of deviations (sample size 177)

Deviation	Frequency (%)
Both technical and operational	12
Only technical deviation	20
Only operational deviation	43
No deviation	11
Unclear	14
Total	100

Technical deviations

In 57 accidents (32%), technical deviations in the control system were found. This gave a total of 60 deviations, which are shown in Table 3. In 24 more cases a technical deviation was plausible. "Functional change" means change of working mode of the control system by, for instance, a switch. Almost all these cases are presses, where the two-hand security device was decoupled through a key.

TABLE 3

Technical deviations in the control system at 56 accidents

Deviation	Number
Technical fault	33
Functional change	14
Safety function disabled	7
Through other equipment	4
Unclear	3
Total	61

The machine guard status was seldom reported, but cases which were noted included some with apertures wide enough to put in a hand. Another was a press, with a light curtain used for protection and control of the press. However, the space between the curtain and the press was wide enough to leave the operator unprotected when standing close to the press.

20 technical faults on the component level were identified from the data.

TABLE 4

Summary of technical faults

Component		Number
Electrical		24
Circuit breaker	10	
Measuring device	3	
Control device	5	
Other	6	
Mechanical		12
Total		36

When complemented by faults from the investigations by the Labour Inspectorate, a total of 36 faults can be described. These are summarized in Table 4.

Operational deviations

In 98 cases a human error in connection with a control system was identified, all except one caused an inadvertent start. These are given in Table 5. The largest group is unintentional contact with a control switch. This indicates a bad design of control panels.

TABLE 5

Operational deviations in connection with the control system at 98 accidents

Deviations	Number
Unintentional contact with start device	29
Unintentional contact with other, e.g. switch	17
Erroneous movement in repetitive work	8
Mistake at manoeuvre	14
Unknown function to operator	14
Did not know other person was in danger zone	12
Unclear	4
Total	98

Another type of unintentional influence on the control system can be by touching a limit switch, or breaking the light beam to a photocell. To this group belongs a number of cases in which the operator has handled material and thus caused the start.

In 20 cases the machine was started by a person other than the injured one. The most common reason was that the operator did not know that another person was in the danger zone when he started the equipment.

Danger zone

To be in the danger zone has not been regarded as a deviation in itself. But it is of interest to know what the injured person was doing. A summary of the tasks of the persons injured in the danger zone is given in Table 6. Many reasons have been given although in 8% of the cases the task could not be classified. In 5% of the accidents the injured person had no task in connection with the equipment. Of those, 3% occurred in an area not normally regarded as a danger zone.

TABLE 6

Tasks in the danger zone (percentage of all accidents, sample size 177)

Task	Frequency (%)
Change of tool	7
Adjust equipment	19
Material, normal handling	10
Material, correction	23
Material, unclear handling	11
Repair, etc.	12
Cleaning	5
Other	5
Task unknown	8
Total	100

Types of equipment

Different types of equipment are represented among the cases. The most frequent type is presses (32%), followed by transportation equipment (16%), cutting machines, such as saws and lathes, (15%) and packing machines (8%).

For the most common types of machines, the occurrence of some deviations is shown in Table 7. The descriptions of the deviations are abbreviated. They correspond to Tables 3 and 5, and to the text above.

The occurrence of deviations varies for different kinds of equipment. Unintentional contact with a start device is common at presses and cutting machines. The cause of an inadvertent start for transport systems is more often a technical fault, unintentional contact with a limit switch, or breaking the beam to a photocell.

Industrial robots

Accidents with industrial robots were studied separately. 29 such accidents were found between 1979 and 1981. This figure is too low, as a

TABLE 7

Frequency of deviations per accident for different types of equipment

Deviation	Equipment			
	Press (%)	Transport (%)	Cutting (%)	Total (%)
Technical				
Technical fault	18	28	27	19
Functional change	20	3	4	8
Safety function disabled	5	0	4	4
Through other equipment	4	10	0	3
Operational				
Unintentional, start device	30	3	19	16
Unintentional, e.g. switch	0	17	4	8
In repetitive work	20	0	8	10
Mistake at manoeuvre	4	3	4	5
Unknown function	7	7	23	8
Did not know other person was in danger zone	5	3	0	7
Number of cases	56	29	26	177

simple check showed. Five robot accidents were found by direct contacts with a few companies. Of these, two were not classified as being robots.

An earlier estimate (Carlsson et al., 1979) suggested 2.5 accidents per 100 robots per year. In this study we obtain a lower frequency — 1 accident per 100 robots per year. We have assumed a mean value of 1,000 industrial robots in Sweden. However the frequency is too low due to missing data. The size of the error is difficult to estimate.

A simple classification of the accidents is given in Table 8. Another study has been made of robot accidents by Sugimoto (1977), but with a different classification. Due to lack of sufficient information the results cannot be directly compared in tables. However, there are some notable differences in the results. In this study, 60% of the accidents occurred when the operator made a correction on the robot or the handled material. The

TABLE 8

Distribution of situations in 29 accidents with robots

Situation	Number
Inadvertent start	10
Normally working robot	10
Energy not from robot	6
Manual handling of robot	3
Total	29

corresponding result in Sugimoto's study is 17%, according to our interpretation of his tables. Another of Sugimoto's results is that half of the accidents occurred during programming and testing, while in this study the figure is 14%.

DISCUSSION

The number of accidents in connection with control systems has been estimated at approximately 2,000 per year, corresponding to 2% of the occupational accidents in Sweden.

An initial issue was why a person should be in the danger zone of an automatic machine. Several reasons have been noted, although in only 2% of the cases was it recorded that the injured person had no reason to be there.

Another question was how often interlocks were intentionally disabled. This was noted in 7 cases (4%). The figure is probably larger, because this type of information might be censored. A more common case (8%) is the disabling of safety functions by switching a key. This applies especially for presses (20% of the accidents).

Technical faults were observed in 19% of the accidents. This figure is probably too low. In particular, intermittent faults might be hard to discover. In general it can be concluded that technical faults are a significant cause of accidents, especially for some types of equipment, e.g. transportation systems (28%).

Machine guard status was not studied, due to lack of information, but a general impression was that the operators could work in the danger zone while the machines were in operation.

Operational deviations were identified in 55% of the accidents. Another observation is that, in 68%, no technical deviation was noted, indicating that the control system worked normally. This gives the impression that the systems are tolerant of human errors and of imperfections in predicting the machine movements. The high figure (16%) for a start caused by unintentional contact with a start device is remarkable. However, this is related to the search code, which might favor this type of event.

One observation is that the operator seems to have corrected a deviation of the material or in the equipment in one third of all accidents. Thus, hazards in this kind of work seem to be a major problem.

The correction of deviations is linked to the problem of machine guards, which are obstacles to these corrections. These matters lead to speculation concerning the design process. There may be gaps in the responsibilities of the control system designers, production engineers, safety engineers, etc. This is quite natural because of their different backgrounds and focuses of interest.

There are two main problems connected with the approach to this study:
(1) The quality of information in the reports.
(2) The selection of accidents.

The reports are usually short. This means that some reports are rejected when it is not possible to judge if a control system is involved. This may also affect the classification of an accident, which means that some parameters might be coded as unclear. However, we obtained more information from the reports than we originally expected.

The intention behind the method of selecting accidents was to get a fairly representative picture of accidents involving control systems. The large number of accidents which either have to be rejected or are not coded in the appropriate way, indicates that distortions in the distributions are likely. Certain kinds of accident sequences or types of equipment can be coded in such a way that they will be over- or underrepresented.

However, it can be concluded that the different risks and problems shown in the report exist. They can be more or less important for different kinds of equipment, but the designers and buyers of machines with control systems should be aware of the different risks that might occur.

Another conclusion that can be drawn from this study is that many accidents could be prevented by an improved design of control systems. This will become more important in the future because of increased automation. However, designers need support in this task, so further research is needed both in identifying problems and risks, and in compiling criteria and good praxis for safety aspects of control systems.

To sum up, some points are given on the design of automatic systems:

(1) Check the protection of the danger zone (compare point 2 below).

(2) Deviations and disturbances in material and equipment will occur. Plan for a safe correction.

(3) Automatic systems often have two working modes: Manual and Automatic. The Manual mode must also be carefully designed, and it must be easy to understand its functioning.

(4) Avoid introducing working modes in the control system where safety functions are disabled.

(5) Observe the design and protection of start devices in order to avoid unintentional starts.

(6) Ensure high reliability of the safety functions; limit switches especially have a tendency to fail.

REFERENCES

Andersson, R. and Lagerlöf, E., 1983. Accident data in the new Swedish information system on occupational injuries. Ergonomics 26(1) (1983): 33–42.
Carlsson, J., Harms-Ringdahl, L. and Kjellén, U., 1979. Industrirobotar och arbetsolycksfall. Royal Institute of Technology, Trita — AOG 0004, Stockholm. (English translation: Industrial robots and accidents at work; Trita — AOG 0026, 1983.)
Faxö, B., 1981. The National Board of Occupational Safety and Health, Stockholm, personal communication.
Harms-Ringdahl, L., 1982. Riskanalys vid projektering — Försöksverksamhet vid ett pappersbruk. Royal Institute of Technology, Trita — AOG 0020, Stockholm.

Harms-Ringdahl, L., 1983. Vis av skadan — Försök att systematiskt utreda och förebygga olycksfall vid två pappersbruk. Royal Institute of Technology, Trita — AOG 0025, Stockholm.

Rouhiainen, V., 1982. Inadvertent start causing accidents; J. Occupational Accidents 4(2—4) (1982): 165—170.

Sugimoto, N., 1977. Safety engineering on industrial robots and their draft standard safe requirements. Proc. 7th Int. Symp. on Industrial Robots, Tokyo, 1977.

ABSTRACTS

Human Aspects of Safety in Offshore Maintenance

REIDAR ØSTVIK

The Foundation of Scientific and Industrial Research at the Norwegian Institute of Technology (SINTEF), N-7034 Trondheim-NTH (Norway)

Deficiencies in the maintenance function within an industrial process plant are usually revealed in connection with the practical performance of maintenance jobs. Usually they can be observed in the form of inefficiency, reduced functionability and unsatisfactory fulfillment of formal obligations on the practical work level. However, these effects can frequently be shown to be related to limitations in human functioning and system errors on the different levels of management in the organization. Such tendencies are further developed and "legalized" downwards through the organization, and an informal practice based on local, practical and person-related motives is established. A judgement of the situation based on information from the formal, official system will therefore not give a complete picture of the function of the system.

Potentials and Limitations of Risk and Safety Analysis – Experience with the SCRATCH Program

JANN H. LANGSETH

SINTEF, N-7034 Trondheim-NTH (Norway)

The SCRATCH (Scandinavian Risk Analysis Technology Cooperation) research program is a coordinated program initiated by NORDFORSK. It consists of 33 individual projects carried out in Denmark, Finland, Norway and Sweden. These 33 projects are financed by research councils or industry within each country.

The program is carried out under the leadership of a board and a secretariat within NORDFORSK. Almost 100 people from a number of research institutes have been involved. This group, together with representatives from industry and inspectorates, have participated in seminars and meetings arranged by SCRATCH. The program was set up in 1978 and the final report will be ready in 1984.

The purpose of the program is to systematize experience with the national projects. The aim of this work is to develop methods of analysis and calculation of risk in connection with industrial systems. These methods will be a supplement to existing standards and regulations and empirical risk judgements, since the technical and social development has made these approaches inadequate.

The technical and analytical side of these methods is of course of great value, but that is probably not the most important result. More important is the understanding of how such analysis should be used to make the most of its potential.

Risk analysis is a very powerful tool in achieving increased safety, but has no life of its own. It must fit into the context of known premises. One must thus have a proper understanding of why the analysis is carried out and what decision process the answers shall fit into. This may seem obvious, but in real life it is not.

One aspect of this is presentation of results. Figures like 10^{-6} may seem satisfactory to the analyst, but will often be of no use to the decision maker. The analyst should have a clear understanding of the acceptance criteria used by the decision maker and present his results accordingly.

Risk analysis cannot be successfully carried out without close contact or even participation from the groups that are affected by the results. No analyst has a complete understanding of the system he is studying, and no transfer of results is perfect.

Furthermore, risk analysis is a way of thinking rather than one specific scientific method. It is easy to adjust the problem to fit one's favorite method, but that will probably lead to solving the wrong problem.

Finally, "technical systems" in industry are not purely technical systems. People are involved, carry out their work and they are tied together in an organization. People and organizations are difficult to analyse, at least in the manner technologists are used to analysing things. Nevertheless, classifying accidents as "human error" will not be acceptable in the future as an easy way out. Methods that deal with the whole system, including people and organization of work, must be developed. SCRATCH has put together the scientists needed for this task. The rest is up to them.

Identification of Accident Risks in Maintenance

JOUKO SUOKAS

Technical Research Centre of Finland, Occupational Safety Engineering Laboratory, Box 656, SF-33101 Tampere 10 (Finland)

Maintenance has often been considered too difficult for a safety analysis because of variations and changes in the working environment and work tasks. There are, however, two different ways of identifying potential accident hazards in maintenance: (1) the analysis of maintenance organization and planning of work tasks, and (2) the analysis of repetitive maintenance tasks.

The MORT-method is an efficient way to identify weaknesses and potential oversights in an organization. These are often factors lying behind immediate accident hazards. The MORT-analysis has assisted, e.g., to improve the coordination between maintenance tasks; to define more accurately work tasks and responsibility; to develop work methods and appliances; and to improve safety awareness at all levels in an organization.

Specific maintenance tasks, especially repetitive work tasks, can be analyzed with work safety analysis. The search for accident hazards is based on observation of job performance in various phases and interviews of workers.

Software Safety in Microprocessor-Based Machinery

SØREN LINDSKOV HANSEN

Elektronikcentralen, Venlighedsvej 4, DK-2970 Hørsholm (Denmark)

The microprocessor is rapidly becoming an integrated part in machines and process control equipment. Where safety is concerned this emphasizes the responsibility of the software producer. It is generally agreed that programs are not easily analysed with respect to their safety properties. One of several reasons for this is that software is comparable to a design and is not to be treated as or compared to a traditional electronic or mechanical component. There is some knowledge available as to which constituents are necessary in developing software with high quality safety properties. Within the programming community there is however no established agreement on what constituents are sufficient in order to achieve a defined level of safety.

The five Nordic Health and Safety Executives have addressed these problems by defining a research project within Elektronik Centralen. The project is funded by the Nordic Council of Ministers.

This paper describes the findings of the first phase of the project concentrating on formulation of problems and their associated prophylactic design elements.

Practical Utilization of Safety Analysis Results

J.R. TAYLOR

Risø National Laboratory, Postbox 49, DK-4000 Roskilde (Denmark)

Two full-scale quantitative safety analyses of chemical plants are described. The analyses can obviously be used as a basis for decisions on design improvement. But a naive use leaves an enormous potential for safety improvement untapped. By interpreting the safety analysis as a basis for safety management, a wider range of objectives can be achieved such as (1) defence against as yet unidentified hazards; (2) setting priorities and time scales for improvement; (3) distinguishing between trivial and serious operational disturbances; (4) achieving a better relationship between authority requirements and practical safety measures; and (5) providing a point of focus for safety campaigns.

The cost of a safety analysis intended to fulfill these goals is about 30% higher than for a conventional risk analysis, and requires a much more flexible range of safety criteria.

The Worker as a Safety Resource in Modern Production Systems

JAN HOVDEN and TERJE STEN

SINTEF, Safety and Reliability Section, N-7034 Trondheim-NTH (Norway)

The traditional focus on human error, accident causation, and risk conditions in safety research should benefit from adding considerations of the worker as a safety resource and problem-solver. Identification of undesired risk behaviour, risk conditions and accidents is no guarantee of obtaining relevant knowledge of safe behaviour and safety measures.

In a research project at SINTEF on man–machine communication studies on operators' monitoring and crisis intervention of sub-sea oil production systems, the theoretical approaches and research strategies stress the positive concept of man; e.g. coping, performance, skills, adequate handling of non-programmed tasks, etc., as a basis for recommendations on design of technical systems, work organization, and training.

Accidents and Variance Control

GORDON ROBINSON

University of Wisconsin-Madison, Department of Industrial Engineering, 1513 University Avenue, Madison, WI 53706 (U.S.A.)

New forms of work, in new organization designs, are placing maintenance tasks -- in the form of sociotechnical system's "variance control" -- in the hands of many more workers. This could have serious implications for a safety program. The function of variance control is discussed in relation to production and maintenance and as it has developed within sociotechnical systems theory and practice. Newer organizational designs that present further problems are presented. Design issues are then discussed, including those in the technical system (including the workplace) and in the social system (including training and responsibility).

Safety Considerations in the Design of Factories – A Study of Three Cases

GRETA FÅNG and LARS HARMS-RINGDAHL

Occupational Accident Research Unit, Royal Institute of Technology, S-10044 Stockholm (Sweden)

The design procedure of three factories has been studied. The safety considerations were of special interest. Also, the treatment of other factors of importance to the work environment was studied. One purpose was to get an idea of the current practice of preventing accidents in the design stage. The second aim was to find potential improvements and obstacles to them. General decisions, time schedule and economy are important factors.

In general, safety matters receive only limited attention during the design procedure. Safety is mainly cared for by safety inspections during and after the installation, resulting in machine guards and service platforms. Noise and ventilation problems have received most attention.

Another general observation is that the time schedule was very tight, but could be fulfilled. This means that little time was available for modifications and analysis of accident risks, if this was the intention. However the cases had a pre-history of several years. This indicates that the time could be used in a better way.

CONCLUSIONS

OCCUPATIONAL ACCIDENT RESEARCH: WHERE HAVE WE BEEN AND WHERE ARE WE GOING?

JERRY L. PURSWELL

University of Oklahoma, School of Industrial Engineering, 202 W. Boyd, Suite 124, Norman, Oklahoma 73019 (U.S.A.)

and KÅRE RUMAR

National Swedish Road and Traffic Research Institute, S-581 01 Linköping (Sweden)

ABSTRACT

Purswell, J.L. and Rumar, K., 1984. Occupational accident research: where have we been and where are we going? *Journal of Occupational Accidents*, 6: 219—228.

The purpose of this paper is to try to evaluate and highlight the International Seminar on Occupational Accident Research held in Saltsjöbaden, Sweden, in 1983. Initially a background is given in terms of the development of studies of occupational accidents in recent decades. Accident statistics — the basis for all accident studies — are discussed in relation to both a previous seminar, held in Sweden in 1975, and to the future.

The question of whether occupational safety should try to find one model, a uniform methodology, is discussed but not resolved. The problems of validity, reliability and countermeasure evaluation deserve and have received considerable attention during the seminar. The most obvious change from the 1975 seminar was found to be the proportion of authors concentrating on finding effective countermeasures. In the near future we shall probably see further research to guide employers and lawmakers in their efforts to change the working environment, more standardization in terminology and statistics, a convergence concerning methods used, more interest in cost effectiveness studies and a further closing of the gap between research and application.

INTRODUCTION

It is a distinct pleasure for the authors to attempt to evaluate the results of the International Seminar on Occupational Accident Research, while at the same time it is a difficult assignment. This seminar took place eight years after a seminar on the same topic was convened in Stockholm (Swedish Work Environment Fund, 1976), so it is useful to refer to the results of that seminar in this paper from time to time to draw various comparisons.

The decade of the seventies dawned with the perception in many industrialized countries that injury rates were at unacceptably high levels and fresh action would be required of employers, governments and workers to reduce these injury rates. Although there were varying levels of national consensus in the industrialized countries on what should be done to reduce the injury rates, there was a general perception that occupational accident research should be directed to the generation of new information, and in turn new countermeasures for reducing occupational injury rates. Some of this motivation for more research was due to the successes which had occurred in reducing injuries from traffic accidents during the sixties through research in these same countries.

With few exceptions, the basic approach to understanding occupational accidents in the industrialized countries had not changed for the 15 years prior to 1970. Heinrich's (1959) text had focused most of the attention on worker behavior (unsafe acts), and the countermeasures for reducing injuries were in turn focused on better training, education and motivation for workers. Thus, this approach resulted in a limited set of possible countermeasures for reducing workplace injuries.

There is always a lot to be learned from history. This is true also for occupational safety. Therefore let us take a look at the history of work (Rumar, 1982). Not too long ago production was synonymous with handicraft. We had our fingers on what we were producing. It was a small selfpacing system with built-in feedback. But efficiency was low. In order to improve efficiency man developed independent energy sources -- engines (steam, explosion, electrical). By means of these engines the efficiency of production increased tremendously. But at the same time feedback was

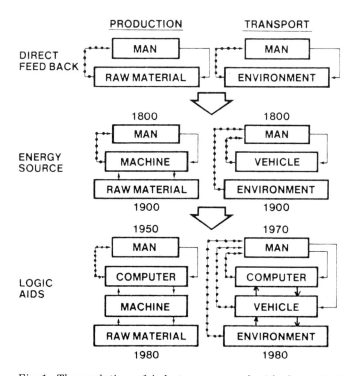

Fig. 1. The evolution of industry compared with the evolution of transport to a man—machine system. The introduction of an energy source and computers have gradually relieved man of many tasks but at the same time introduced new problems.

reduced, mental tasks like attention have increased due to higher speed, but selfpacing has disappeared and job satisfaction has decreased. This process is outlined in Fig. 1.

The process has given us a system in which technical development has been enormous during the last 100 years. The human link in the system has hardly changed at all. It has several built-in limitations that we have to take into consideration, but it is a system the function and safety of which can be changed primarily by changing the interaction and not one component, e.g. man.

As a result of this development, occupational accident investigators in different countries began to search for broader, more acceptable explanations for the causes of accidents and injuries and, in turn, to expand their efforts with regard to collection and countermeasures development during the seventies. Researchers from the disciplines of engineering, psychology, sociology, medicine, industrial hygiene and other areas all began to contribute ideas and research methodologies from their respective disciplines. Different disciplines have their own concepts of acceptable research methods and countermeasure development based on their research findings. This diversity has necessarily created some problems in defining a unique "occupational accident research methodology" which could be widely accepted among researchers and practitioners in the field. The discussion which follows focuses on some consequences of this diversity in data collection and countermeasure development, both bad and good, which become apparent in the different papers presented during this seminar. An attempt is then made to draw some conclusions (or perspectives) about the future of occupational accident research.

ACCIDENT STATISTICS

An important starting point for a seminar on occupational accident research would appear to be types of statistics collected by governments as a measure of safety and health in the workplace. Several papers in this seminar addressed this problem, either directly or indirectly in terms of data availability and its usefulness for countermeasure development. The experience in developing and using data registers which has been gained since the seminar of 1975 was apparent in the papers presented by authors from several countries. In some cases the data collected were sufficiently detailed to suggest possible countermeasures, while in other instances they could only suggest areas where further research was needed before countermeasures could be developed.

An appeal was made during this seminar for more effort to standardize the definitions used as well as the basic data collected in order to permit more useful comparisons from one country to another. Such basic indicators of safety performance as injury incidence rate or injury severity rate often cannot be rationalized from one country to another.

The International Social Security Administration (ISSA) and the International Labour Office (ILO), Geneva, have tried unsuccessfully for many years to persuade member states to standardize their collection and reporting of basic occupational injury statistics. It is therefore not surprising that safety researchers have not succeeded in bringing about some standardization of official statistics in their respective countries. However, it is still possible for occupational safety researchers to work to standardize various definitions and research terminology. Success in this endeavor could significantly enhance the exchange of ideas among researchers and permit a cross-fertilization of ideas. Future seminars should perhaps highlight this objective on the agenda.

It seems important here to stress that there are complementary methods to accident statistics as regards diagnosis of safety situations, measuring the effect of safety measures, analysis of accident "causes". No method has higher validity but most methods have higher reliability — e.g. accident in-depth (case) studies, near-accident investigations, field and laboratory experiments, behavioral observations, interviews, simulations, etc.

OCCUPATIONAL ACCIDENT CAUSATION

If one is to generate countermeasures to prevent accidents and injuries, then there must be some way of relating the "cause" of the accident to the proposed countermeasure. Owners of enterprises are not willing to invest money in modifying equipment, training workers, etc. unless there is a way of demonstrating that the investment is likely to reduce the incidence of injury in the workplace. However, it may be hoped that injury accidents are rare events for most organizational sub-units, which immediately raises the problem of collecting and aggregating sufficient data from many sub-units over time to be able to draw statistical inferences about the "cause" of the injury accident. This need requires the researchers to come to grips with a methodology for data collection or, put in the terms of several papers in the 1975 seminar as well as of the present seminar, a "model" of the accident process which can be used to organize the data collection after an injury accident.*

In his paper "Future trends in accident research in European countries" presented in this seminar, Singleton argues that more emphasis should be placed on the identification of key parameters which can alter the probability of accidents generally, and less emphasis in developing theories or models to explain the cause of *particular* accidents (Singleton, 1984). He calls this a "process oriented" approach and suggests that theoretical studies of the

---

*There are various interpretations of the terms "accident" or "injury accident". See for example Heinrich et al. (1980) for a discussion of the relationships between accidents, minor injury or disabling injury.

systems type directed towards methods of accident control are what is needed in the future.

The concept of analysing the accident process as a flow or pattern of causally and chronologically related deviations was presented by Kjellén in this seminar (Kjellén, 1984). This approach has some attractive features:

(1) It focuses attention on operationally defined parameters of an ongoing industrial process and, in turn, on the consequences of deviations in the parameters beyond specified limits.

(2) It has the ability to incorporate the perspectives of various disciplines regarding the acceptable norms necessary for safe operation.

(3) It has the potential for application to a wide range of production systems and accident types.

Kjellén also notes that, even though the approach or "model" has much to recommend it, there is some question of whether it will be possible to identify general relations between the type of deviation and the risk of accidents.

Perhaps the most important *and* critical aspect of the deviation approach in accident causation is the classification of deviations. Kjellén discusses the taxonomies available for classifying deviations, the more important of which are as follows:

(1) The process model, with time as the basic variable.

(2) The systems model, with a focus on components, subsystems and their interaction in the overall system.

(3) The man-machine-environment model, with a focus on systems ergonomics.

(4) The human error model, with a focus on the sources, occurrence and control of human errors in systems.

(5) The consequences model, which focuses on the severity of deviations in a system parameter.

(6) The management/organizational systems model, which focuses on management control on the overall process and the results of supervision in terms of system safety.

(7) The epidemiological model, which focuses on the boundaries and interactions between host, agent and environment, either in a descriptive or analytical approach.

It is clear that the taxonomies suggested are not mutually exclusive, and different applications will require a combination of approaches.

In a similar approach, Robinson suggests that attention be focused on critical variations in certain safety-related parameters of an industrial process (Robinson, 1984).

Many of the papers presented during this seminar could be classified as examples of one of the taxonomies Kjellén has suggested for classifying deviations in his approach, i.e. the papers of Leplat (1984), Saari (1984), Bengtsson (1984) and Coleman (1984).

Benner (1984) demonstrated in his paper the problems created by inadequate approaches to accident investigation in several agencies of the U.S. government. He argues that a new approach is needed for developing "building blocks" for accident investigations, i.e. an "event" which is characterized by one actor plus one action. These "events" would then be used as the basis for data collection to develop information in terms of event chains or sequences, fault-tree analysis, time/loss analysis and causal factors charting. Benner suggests that safety research has borrowed methodologies from other disciplines in the past, rather than creating its own methodology. He proposes his approach as a methodology in its own right and suggests that a high priority be given to developing a single accident description which would serve all users.

We are then brought back to the basic question of whether there is indeed a suitable approach or "model" for accident data collection, and further whether the "model" or methodology has been borrowed from other disciplines or is unique to the field of safety research. No universally acceptable approach seems to have yet emerged which is unique to occupational accident research, but much progress can be observed since the seminar of 1975. It is also clear that some researchers in the field do not necessarily embrace the concept of a universal approach for describing accidents, but consider the diversity to be acceptable, or in some cases even helpful because of the new ideas generated.

One of the apparent dangers of trying to obtain too much uniformity in the methodology of accident investigation is the prospect of the model driving the problem definition, rather than the problem generating the appropriate model to be used. Thus, there is the tension between data collection for each accident by following a universal "model" versus a more narrowly defined problem and data set. Perhaps there is some insight to be gained at this point from the experience of road researchers.

In several countries, there is a methodology for collecting road accident data that is called bi-level or tri-level, meaning that a minimum data set is collected for all accidents (usually by the police), and additional data is collected by a trained group of investigators for a specific subset of accidents to suggest solutions for a narrowly defined problem area such as collisions with road signs, driving in adverse weather, etc.

Enough data are collected either to suggest countermeasures directly or to define a research program and the researchers then undertake a new study of another narrowly defined problem. Some of the papers (Cohen, 1984, for example) illustrate the application of this approach to occupational safety. It is the authors' opinion that this approach could be used more widely in occupational safety research.

There seems to be a tendency among occupational accident researches to stick too long with the accident event itself. As was often mentioned during the seminar, accident statistics have their limitations — one of the main ones being that it is hard to get an understanding of what happened and why through statistics. But once the problem is established, there are good reasons to look around for other methods to find an effective countermeasure.

VALIDITY, RELIABILITY AND EVALUATION

If one compares the seminar of 1975 with this seminar, it is clear that more of the present seminar was spent in addressing questions of validity and reliability in the methods of occupational accident research. This concern is certainly appropriate because of the wide variations sometimes found to exist in the way different researchers will evaluate and record the "facts" of the same accident, even when they are supposedly following the same methodology and work in the same discipline. When this occurs, there is a loss of credibility and the possibility that countermeasure efforts will be misdirected.

Safety research efforts are still hampered by the difficulties which exist in obtaining valid measures of exposure. In the absence of some method for normalizing or rationalizing the incidence and/or severity data collected, there is a problem in assigning correct priorities for countermeasure development. Even though some progress can be observed since the seminar of 1975, many of the studies reported during this seminar contained little or no exposure data. Since employers and government policy-makers always have competing demands for available resources, a prioritization scheme is necessary for implementing various safety and health measures.

Some papers in this seminar have addressed the evaluation of various safety measures (Hale, 1984; Nilsson, 1984, for example). As more effort is directed to improving workplace safety, it is increasingly necessary to build an evaluation component into the countermeasures implemented.

Evaluation is frequently attempted by comparing statistical measures of safety before and after the countermeasure program was implemented. This approach is selected because of the difficulty in employing other more desirable alternatives such as control versus experimental group comparisons. As Nilsson (1984) points out, the before-versus-after comparisons may be in error because the results observed may merely represent regression-to-the-mean effects, rather than the effects of the countermeasure being evaluated.

Future seminars should perhaps give more attention to the problems of evaluation.

IMPLEMENTATION OF COUNTERMEASURES

The most striking difference between the seminar of 1975 and this seminar is the number of papers presenting a description of the implementation of some countermeasure to address a problem of workplace safety. This is no doubt a healthy sign of a maturing research effort, with many different countermeasures being launched to deal with specific problems identified through research studies.

The paper by Komaki (1984) presents an excellent example of implementing a successful behavioral modification program to reduce injuries in a firm. She successfully implemented a positive feedback program for safe work behavior, overcoming the usually accepted truism that there are no perceived positive rewards for safe work behavior, only negative ones for unsafe work behavior.

Another example of behavior modification/reduced risk-taking was described in the paper by Sundström-Frisk (1984). The very important influence of wage payment schemes on worker safety was analysed for the forest industry, as well as the effects of introducing safer equipment. Suspension of a piece-work wage payment system in favor of a salary payment program was shown to produce a marked reduction in both the incidence and severity of injuries to forest workers. This result was in addition to a reduction of injury rates produced through the introduction of safer equipment. Identifying and removing the incentive for risk-taking behavior has been a concern of safety researchers for many years (Rockwell, 1961; Lagerlöf, 1977; Rowe, 1977, for example). It seems likely that this avenue of research will remain a productive one for future safety researchers.

FUTURE SAFETY RESEARCH

The following observations are offered about the future of safety research, based both on the papers presented during the seminar and on the authors' collective experience in the research field for many years.

(1) The seventies produced new legislation in many of the industrialized countries of the West to improve safety at the workplace. A major effort to improve workplace safety through the enforcement of rules and regulations, training of workers and employers, and research programs to understand accident causes has produced results. Injury incidence rates and fatality rates have declined during the decade. (See Singleton, 1984, for example.)

(2) There is a general conception in several industrialized countries that the next major impact on injury incidence and fatality rates must come primarily as a result of research programs which help employers and government policy-makers to design further countermeasures that are technologically and economically feasible. Stated another way, there is a definite skepticism regarding further improvements in injury and fatality incidence rates as a result of more intensive efforts by factory inspectorates.

(3) The exchange of data and ideas among safety researchers in different countries is definitely a helpful exercise. This exchange could be facilitated if more standardization could be achieved in terminology and methods of computing safety statistics.

(4) The methodology for performing safety research has become much better understood over the last decade. While there is not yet a universally

acceptable method, there is convergence in the number of methods being used. The advantages and disadvantages of different approaches are better understood by researchers as a result of meetings such as this seminar.

(5) The recent worldwide recession has had the effect of reducing funds available for safety countermeasures in many countries. This increased pressure on cost effectiveness has made the question of evaluation of safety countermeasure more urgent. This seminar has laid a foundation on which more emphasis could be given to this area in future meetings.

(6) There is a growing capability among safety researchers in their ability to bridge the gaps often found between research and application. Safety practice for the first 60—70 years of this century was based primarily on a few well-worn "principles" with little new information generated through research. There is now a much greater awareness of the need for research and a willingness by researchers to develop countermeasures based on their research results.

REFERENCES

Bengtsson, B., 1984. Epidemiology of occupational accidents. J. Occupational Accidents, Vol. 6, No. 1/2, Proc. Int. Seminar on Occupational Accident Research, Saltsjöbaden, September 1983.
Benner, L., 1984. Accident models: How underlying differences affect workplace safety. J. Occupational Accidents, Vol. 6, No. 1/2, Proc. Int. Seminar on Occupational Accident Research, Saltsjöbaden, September 1983.
Cohen, H.H., 1984. Advantages and limitations of various methods used to study occupational fall accident patterns. J. Occupational Accidents, Vol. 6, No. 1/2, Proc. Int. Seminar on Occupational Accident Research, Saltsjöbaden, September 1983.
Coleman, P.J., 1984. Descriptive epidemiology in job injury surveillance. J. Occupational Accidents, Vol. 6, No. 1/2, Proc. Int. Seminar on Occupational Accident Research, Saltsjöbaden, September 1983.
Hale, A.R., 1984. Is safety training worthwhile? J. Occupational Accidents, Vol. 6, No. 1/2, Proc. Int. Seminar on Occupational Accident Research, Saltsjöbaden, September 1983.
Heinrich, H.W., 1959. Industrial Accident Prevention. McGraw-Hill, New York, NY.
Heinrich, H.W., Petersen, D. and Roos, N., 1980. Industrial Accident Prevention. McGraw-Hill, New York, NY.
Kjellén, U., 1984. The role of deviations in accident causation. J. Occupational Accidents, Vol. 6, No. 1/2, Proc. Int. Seminar on Occupational Accident Research, Saltsjöbaden, September 1983.
Komaki, J.L., 1984. A behavioral approach to work motivation. J. Occupational Accidents, Vol. 6, No. 1/2, Proc. Int. Seminar on Occupational Accident Research, Saltsjöbaden, September 1983.
Lagerlöf, E., 1977. Risk identification, risk consciousness, and work organization — three concepts. In: Swedish Work Environment Fund, Proc. French-Swedish Symp. on Research on Occupational Accident, Stockholm, 1976.
Leplat, J., 1984. Occupational accident research and the system approach. J. Occupational Accidents, Vol. 6, No. 1/2, Proc. Int. Seminar on Occupational Accident Research, Saltsjöbaden, September 1983.

Nilsson, G., 1984. A review of the traffic safety situation in Sweden with regard to different strategies and methods of evaluating traffic measures. J. Occupational Accidents, Vol. 6, No. 1/2, Proc. Int. Seminar on Occupational Accident Research, Saltsjöbaden, September 1983.

Rockwell, T.H., Galbraith, F.D. and Center, D.H., 1961. Risk-acceptance research in man—machine systems. Ohio State University, Engineering Experimental Station, Bulletin No. 187, Columbus.

Rowe, W.D., 1977. An Anatomy of Risk. John Wiley and Sons, New York.

Rumar, K., 1982. The human factor in road safety. Invited paper at the 11th Australian Road Research Board Conference, Melbourne, Australia, Augst 23—27, 1982.

Saari, J., 1984. The disturbance of information flow and accidents. J. Occupational Accidents, Vol. 6, No. 1/2, Proc. Int. Seminar on Occupational Accident Research, Saltsjöbaden, September 1983.

Singleton, W.T., 1984. Future trends in accident research in European countries. J. Occupational Accidents, Vol. 6, No. 1/2, Proc. Int. Seminar on Occupational Accident Research, Saltsjöbaden, September 1983.

Sundström-Frisk, C., 1984. Behavioural control through pay systems. J. Occupational Accidents, Vol. 6, No. 1/2, Proc. Int. Seminar on Occupational Accident Research, Saltsjöbaden, September 1983.

Swedish Work Environment Fund, 1976. Proc. Seminar on Occupational Accident Research, Stockholm, Sweden, 1975.